IEEE ENGINEERS GUIDE TO BUSINESS

WRITING FOR CAREER GROWTH

by
David L. McKown, PE

President
InnerLink Publication Services, Inc.
Pitcairn, PA

**The Institute of Electrical and Electronics Engineers, Inc.
New York, New York**

IEEE *Engineers Guide To Business Series*

Editor: Barbara Coburn
Cover Design: Ralph Ayers

<u>1993 EDUCATIONAL ACTIVITIES BOARD</u>

Vice President
Edward A. Parrish

R. Nichols	K. Laker	A. Van Gelder
D. Hodge	V. K. Bhargava	C. Sechrist
W. Carroll	D. Jackson	D. Conner
P. Grosewald	F. Dill	M. Van Valkenburg
M. Masten	L. Feisel	J. Yeargen
M. Andrews	B. Eisenstein	

Copyright © 1992 by
THE INSTITUTE OF ELECTRICAL AND ELECTRONICS
ENGINEERS, INC.
345 East 47th Street, New York, NY 10017-2394

*All rights reserved. No part of this book may be reproduced,
in any form nor may it be stored in a retrieval system or transmitted
in any form without written permission from the publisher.*

Printed in the United States of America

Library of Congress Cataloging-in-Publication Data

McKown, David L., 1947-
    Writing for career growth / David L. McKown.
      p.   cm. -- (IEEE engineer's guide to business ; v. 2)
    ISBN 0-7803-0304-0 : $19.95
    1. Business writing. 2. Business communication. I. Title.
II. Series.
HF5718.3.M365 1991
808'.0662--dc20
                                                                   91-42182
                                                                       CIP

Foreword

A number of years ago, I was approached by a low-keyed graduate student in business who wanted to take a course in drama. Prior to entering business school, this individual had earned two degrees in engineering, and had been a practicing engineer for more than six years. As you might imagine, the course selection caught me by surprise. Why a course in drama, especially for such a somber individual?

The fact of the matter is that this engineer had recognized the importance of communication in practice -- that is, being able to convey, either in person or in writing, important ideas and issues. Good communication is what underlies the transfer and acceptance of the analysis and recommendations of employees. This student had recognized this, and wanted to develop the skill of selling his ideas.

Similarly, good written communication is essential for selling ideas and knowledge within an organization. Such communication often is more difficult when technical analysis underlies the issues being presented. The best ideas still require persuasive communication to have their full impact. The ability to communicate is critical for transforming analysis and recommendations into action.

Writing for Career Growth is directed at making engineers more effective in their practice. It is an important document for helping the technical professional communicate in a meaningful way. It is written in a way to provide rapid guidance to writers, and it does this in an interesting and insightful manner. I am sure many technical professionals will benefit considerably from *Writing for Career Growth*.

Dr. Robert S. Sullivan, Dean

Graduate School of Industrial Administration
Carnegie-Mellon University
Pittsburgh, Pennsylvania

PREFACE

"Oh, no! Now you want to tell me how to write!? Wasn't Advanced Calculus enough? Look, I just want to be left alone to do what I do best - engineering. Let me take care of the important technical work, and get some English major to do the reports. Anyway, isn't that what the Tech Pubs department is for?"

Well, I guess I've heard that before. In fact, I've said it! It does seem odd that you spend years of your life learning how to be a good engineer, and then someone comes along and says it's not enough, that you've got to "learn to communicate" for Pete's sake. But I found that it's true. It isn't enough to be a good engineer if you want to be happy and successful in your career. Why? Because, unless you are independently wealthy, have your own lab, and work alone, you must depend on others some time. That means you'll have to communicate (that word, again) with them, and that often means "put it in writing."

This book is meant to help you do that (put it in writing) quickly and effectively, so that others get the message you intend and so that you can get back to engineering. For the new (or newer) engineer, this is particularly important, because *you will be remembered by what you write as much as what you do.* In many cases, your superiors will know you only by the progress reports you write, or by the snippets of them that your boss includes in the monthly report. Your coworkers may know you mainly through memos you write. Your professional peers across the country may know you only by the papers you publish in professional journals. In short, to many people, you *are* what you write ... and if it is perceived to be unprofessional, so are you.

ACKNOWLEDGEMENTS	Thanks in the highest go to Joan Nagle and Lois Taylor whose review and editing of the manuscript was invaluable and saved me from myself in many instances. Also, Gladys, my long-suffering spouse, deserves special mention for her sound counsel and encouragement, and for putting up with the countless hours I spent staring at the computer screen and the thousand promises I made and then broke ("Of course, dear; just a few more hours and it'll be done!").
ABOUT THE AUTHOR	David L. McKown is a licensed Professional Engineer with degrees in electrical engineering, bioengineering, and business. He learned the hard way that effective written communication is essential to career growth during 15 years in technical engineering, marketing, and sales, and another 10 in technical publishing. In 1991, propelled by the dissolution of his department within a major corporation, he formed his own company, The Broadway Group, Inc., to continue to offer full service publishing to high technology firms. The company employs 15 professional editors, writers, designers, and illustrators near Pittsburgh. The motto of The Broadway Group is *We Document Technology!*
COLOPHON	This book was produced using the following tools:

Software

Initial outline:	Grandview®, v. 2.0
Word Processors:	Amí Pro©, v. 1.2b
	Interleaf Publisher©, v. 1.01
Illustrations:	Amí Pro©, v. 1.2b
	Interleaf Publisher©, v. 1.01
Tables:	Amí Pro©, v. 1.2b
Page formatting:	Amí Pro©, v. 1.2b

Hardware

Computer:	Northgate Elegance© 386/33
Monitor:	Nanao 9070S©
Printer:	Hewlett-Packard LaserJet III©
Scanner:	Hewlett-Packard ScanJet Plus©

Typography

Body type is set in Times Roman 10/12
Side heads are set in Helvetica 12/12
Chapter Heads are set in Helvetica 14/17
Table body text is set in Times Roman 8/9, tight kerning
Table heads are set in Times Roman 8/9

Contents

Part I: Getting Ready to Write

Chapter 1: What is Written Communication?
Why is Written Communication Important? 1-2
How Can I Learn About Written Communication? 1-2

Chapter 2: The Importance of Written Communication
Write for Success 2-1
Write for Efficiency 2-2
Write for a Reason 2-2
Getting Started 2-3

Chapter 3: Define Your Audience
Audience Types 3-1
Audience Complexity 3-5
Audience Needs 3-6
Language Components 3-7
Case Study 3-9
Checklist for Audience Analysis 3-16

Chapter 4: Define Your Purpose
Why are you Writing? 4-1
Reasons 4-2
Checklist for Writer Analysis 4-11

Chapter 5: Get Organized
Organization and Understanding 5-1
    Organizing and the Audience 5-1
    Organizing and the Author 5-1
Approaches to Organization 5-2
    Clumping 5-2
    Tell 'Em Three Times 5-4
    Cause and Effect 5-5
Organization of Types of Documents 5-5
    Report 5-5
    Proposal 5-5
    Manual 5-5
    Tutorial 5-6
Techniques of Organization 5-6
    Outlining 5-6
    Writing Inside-Out 5-8
Visual Clues 5-10
    Parts of a Document 5-10
    Typography and Page Design 5-11
    Other Reader Aids 5-14
Summary 5-15

Chapter 6: Include Supporting Data

- Clarity 6-1
- Authority 6-1
- Visual Elements 6-1
  - Tables 6-2
  - Lists 6-2
  - Illustrations 6-5
    - Charts and Graphs 6-5
    - Sketches 6-10
    - Drafting Drawings 6-10
    - Technical Illustrations 6-11
- Equations 6-12
- References 6-12

Part II: Writing

Chapter 7: Follow the Rules: Part 1

- Grammar 7-1
  - Parts of Speech 7-3
    - Nouns 7-4
    - Pronouns 7-6
    - Verbs 7-6
    - Adjectives 7-10
    - Adverbs 7-10
    - Prepositions 7-11
    - Conjunctions 7-11
  - Sentences 7-13
    - Sentence Structure 7-14
      - Subject 7-14
      - Predicate 7-14
      - Clause 7-15
      - Phrase 7-15
    - Sentence Types 7-16
    - Bad Welds and Broken Tools 7-16
    - Consistency 7-17
      - Shifts 7-17
      - Danglers 7-17

Chapter 8: Follow the Rules: Part 2

- Vocabulary 8-1
  - Simplicity 8-1
  - Precision 8-2
  - Economy 8-3
  - Context 8-5
  - Gender Neutral 8-6
  - Word Use 8-7

Contents

- Punctuation ... 8-9
  - Period ... 8-10
  - Colon ... 8-11
  - Semicolon ... 8-11
  - Comma ... 8-12
  - Dash ... 8-14
  - Ellipsis ... 8-14
  - Quotation Marks ... 8-15
  - Apostrophe ... 8-16
  - Hyphen ... 8-17
  - Parenthesis ... 8-18
  - Brackets ... 8-18
  - Exclamation Mark ... 8-19
  - Question Mark ... 8-19
- Spelling ... 8-19

Chapter 9: Be Clear

- Hidden Verbs ... 9-1
- Invented Words ... 9-2
- Using Numbers ... 9-3
  - Cardinal ... 9-3
  - Ordinal ... 9-3
  - Fractions ... 9-4
  - Temperature ... 9-4
- Parallelism ... 9-4
- Precision ... 9-5
  - Specificity ... 9-5
  - Redundancy ... 9-5
  - Separation ... 9-6

Chapter 10: Pay Attention to the Fine Points

- Etiquette ... 10-1
- Proprietary Information ... 10-3
  - Copyrights ... 10-4
  - Trademarks and Servicemarks ... 10-6
  - Patents ... 10-6
  - Trade Secrets ... 10-7

Part III: Helpers

Chapter 11: Get the Computer to Help

- Writing Assistants ... 11-1
  - Outliner ... 11-1
  - Thesaurus ... 11-3
  - Thought Processor ... 11-3
  - Hypertext and Hypermedia ... 11-4

CD-ROM Reference 11-5
Editing Assistants 11-5
    Spelling Checker 11-6
    Grammar Checker 11-6
    Redliner 11-7
    Document Comparer 11-7
Managing the Writing Process 11-8
    Filing System 11-8
    Style Sheets 11-9

Chapter 12: Write! . . . .
    Step 1: Define the Job 12-1
    Step 2: Get Ready 12-1
    Step 3: Write 12-3
    Step 4: Rewrite 12-4
    Step 5: Publish 12-5
    Step 6: Relax 12-6

Quick Reference Card

CHAPTER 1   WHAT IS WRITTEN COMMUNICATION?

The subject of this book is *writing for career growth*. Let's discuss that for a moment. Just what sort of writing must an engineer do, anyway? Well, for the most part, it can be described by a phrase that usually brings a chill to the hearts of engineering students (and, unfortunately, most engineers as well): *technical writing*. That does sound formidable, but it is not as grim as it may seem. Technical writing, after all, is merely writing about technical subjects.

And who is better prepared to discuss technical subjects than an engineer? Engineers are trained to understand technical subjects. They are devotees of the scientific method; logic and method are their tools as much as are oscilloscopes and computers. Engineers usually are painstaking in their attention to detail, orderly in their thought, and structured in their logic. These are the same characteristics exhibited in good technical writing. It, too, is detailed, organized, and structured. It, too, follows the scientific method and deals with technical subjects.

In fact, good writing, technical or otherwise, is really the *transfer* of the writer's thought into another's mind. Sounds a little SciFi, but there you have it. (Incidentally, let's not talk about technical writing; let's call it written communication, signifying that the process of thought transference is written and that communication actually does occur.)

To create good written communication, especially about technical subjects, some rules have to be followed, just as rules have to be followed to create a good circuit design or a good stress analysis. Consider this analogy: traffic regulations make it easy and safe for cars and trucks to travel at 55 miles per hour; rules of organization, format, and language make it easy (if not safe) for readers to navigate the freeways of scientific and technical thought. Just as the Highway Patrol impartially enforces those traffic regulations, the engineer should strive for impartiality and objectivity regarding the subject. And, just as drivers use certain techniques to merge into high-speed traffic or to change lanes, the engineer can use a variety of proven techniques for communicating technical information.

You drive to work safely each day because you follow traffic rules and regulations, don't tailgate, use common sense, and have had a lot of experience. In the same way, you can create effective written communication, even for the most technical of subjects. Follow the rules most of the time, maintain your emotional distance, apply common sense all the time, and practice, practice, practice.

WHY IS WRITTEN COMMUNICATION IMPORTANT?

"Oh, yeah. Technical Writing," you say. "I had a course in that once. Back when I was a freshman. Never could understand why an engineer needed all that English stuff." OK, so maybe engineers can become good writers. Maybe even *you* can become a good writer. So what? Why bother?

First, if you did have a course in technical writing as part of your engineering curriculum, you are in the minority. It is a subject that is widely neglected. Second, it *is* important, especially to the newer engineer.

Why? Because success depends not only on your technical excellence, but equally on your ability to communicate effectively in writing to others.

These things are true whether you are a new engineer or an experienced professional:

- It is not enough to have great ideas; you must be able to persuade others that your idea is worth pursuing.
- It is not enough to know what must be done to achieve your goals; you must be able to describe the needed steps to others if you are to work as a team.
- It is not enough to be the project leader; you must let your sponsors and other team members know what you are doing, and accurately describe your progress.

The goal of this book is to provide a set of principles and tools that will help you, the working engineer, be an effective and successful communicator.

HOW CAN I LEARN ABOUT WRITTEN COMMUNICATION?

In *Part I: Getting Ready to Write* you will learn that there are different kinds of technical communications, that each one has a specific purpose, and that you will probably need to use each one at some point in your career (Chapter 2). You will see that it is critical for you to think about who will read your writing, and to know exactly what you want to accomplish (Chapters 3 and 4). And you will be shown how to organize your thoughts and present data (Chapters 5 and 6).

Part II: Writing will show you the impact that style and clarity have on how your writing is perceived by the reader ... and how to make that impact be what you want it to be (Chapters 7, 8, and 9). You'll learn that an important part of writing in business is adherence to the rules of etiquette, propriety, and form, and that paying strict attention to the details helps make the difference between adequate and superior communication (Chapter 10).

Part III: Helpers shows how the ubiquitous computer can be a source of considerable help in improving your writing ... and the dangers of relying on it too much (Chapter 11). Finally, you'll see how all of these parts fit together to form a complete strategy for writing (Chapter 12).

PART 1 : GETTING READY TO WRITE

CHAPTER 2 — THE IMPORTANCE OF WRITTEN COMMUNICATION

As noted in the last chapter, being able to communicate technical information and concepts, clearly, concisely, and in writing is not just important ... it is vital to your success as an engineer.

WRITE FOR SUCCESS

In fact, while good ideas and technical competence are essential to success, it is the ability to communicate clearly that often makes the critical difference. Let's look at this example.

Assume that you have just finished a project that could have a significant effect on the products your company makes. Naturally, you want to let others (particularly your boss) know the results. Look at the following two ways in which the introduction to the report on the results might be written. Which would you rather read? Which would your boss want to pass along to the head of product development?

> *The enclosed report details the findings of the micromanufacturing investigation. Variances of input material and the effect they have on results are discussed. General applications as regards possible implementation and integration into existing product-oriented development efforts are explored.*

or

> *Our experiments with micromanufacturing techniques suggest that we can develop products for existing and future markets. We can now limit the effect of the varying quality of raw material on the finished product.*

The first example is unclear in its purpose and in its message. Is this report good news or bad? Should the head of product application read it or not? Have we wasted our time and money doing this research, or spent them wisely?

The second example comes right to the point: *we can make new products that customers want.* A major production concern is under control.

Readers of technical information want the facts; they want them early and they want them straight. They are busy, and cannot waste time trying to figure out what is meant by a word or a phrase. If you make it difficult for them, you risk misunderstanding, impatience, and annoyance.

Look at it this way: someone is paying for the reader's time. If you waste it, you are wasting (at the least) money, which is not a thing to do if you wish to remain popular. Successful professionals, including engineers, know the value of time and money, and do not knowingly waste either. Engineers plan experiments, design mechanisms, and lay out cities with a view not only to technology but also to economy and the proper use of resources. The same is true for writing.

Good written communication is important to a successful technical career because it helps others realize that you *are* a professional.

WRITE FOR EFFICIENCY

Speaking of the practical equivalence of time and money, have you ever stopped to think about how much time you and your colleagues spend communicating?

This may surprise you, but various sources estimate that engineers and other technical professionals spend from 50 to 70 percent of their time communicating. Further, much of that communication is accomplished through writing. Each day, technical professionals instruct support personnel, tell others about their ideas, enlist help and support, ask for resources, inspire others to get involved, establish leadership and roles, get information, sell a product (or themselves), and accomplish countless other objectives. In fact, in many fields - pure research, for example - a written report may be the only product and the only proof that anything at all has been accomplished.

Think how more productive this reading or writing time could be if everyone practiced the principles of good written communication.

WRITE FOR A REASON

During your career, it is likely you will write for many reasons. Each reason will have its own peculiar demands, require different approaches, target different readers, and force you to think in slightly different ways. In general, technical professionals (and everyone else, except maybe writers of fiction) write within only a few general categories: *to inform*, *to request*, and *to persuade*. (Other authorities describe up to five or six of these categories, but let's keep things simple.) Within these categories, authors write for many reasons, and use many different types of documents to accomplish their ends.

Classic applications of technical writing (manuals, for example) exist to inform, to help others do something. Writing for these documents must be strictly impartial and objective; there is no room for emotion or personal opinion. Reports have a little more leeway, depending on the format. For example, a report may include a Recommendations section, in which the opinion of the author may be required. Even so, that opinion must be based on the facts contained in the report, and not on any personal likes or dislikes. Although all of this might seem obvious, it is far more difficult to do than to say.

Types of documents

Category	Reasons	Document Types
Inform	Instruct	Manual, textbook
	Notify	Announcement, memo
	Warn	Warning, caution, letter
	Interpret	Paper, report
	Clarify	Report, memo
Request	Apply for rights	Patent application
	Get action	Memo, letter
	Obtain information	Memo, letter
Persuade	Make impression	Proposal, newsletter, resume
	Influence decision	Proposal, report, advertisement
	Gain position	Advertisement, proposal, memo
	Gain acceptance	Report, advertisement, paper
	Recommend	Report, memo, paper
	Sell	Advertisement, proposal

GETTING STARTED Many engineers think writing is exactly that: just sit down and start right in. Maybe make a few notes, shuffle them around, but in general just write what you think. And that is why many engineers are poor communicators: there is much more to writing than the writing. It is the successful completion of defining the audience, defining the purpose, and getting organized that makes good written communication of any kind, but especially good technical communication.

The rest of this book concentrates on giving you specific techniques for getting ready to write and for writing. Both are important; neither is sufficient alone.

CHAPTER 3   DEFINE YOUR AUDIENCE

When you sit down to write, you usually have a fair idea of what you want to say. You know what is important, and what you want to happen. However, it may be that what is important to you is not so important to those who will read what you write. So, what really turns out to be important is to think about exactly who will be your readers ... your audience. In fact, at this point it is more important to analyze the audience than it is to worry about the content, because the characteristics of the audience should affect the content, or at least how you write it.

This process of audience analysis involves several steps. These include determining the *audience type*, determining the *audience complexity*, establishing *audience familiarity* with the subject, making sure that you have addressed all the *audience needs*, and choosing the right *language components* to address the audience or audiences you have identified.

A *Checklist for Audience Analysis* appears at the end of this chapter. The entries in it are explained in the following sections.

AUDIENCE TYPES

Few things an engineer writes are read by only one person. Some, like progress reports, may be passed along (if not in their entirety, then in part) to other interested parties, up the management chain, or even to customers. Others, like operating manuals, may be reviewed by several levels of technical experts and management personnel before being delivered to the intended audience, the customer. For our purposes, we will call the person or persons for whom the document is ultimately intended the *primary audience*. Others who may read the document we will call *secondary audiences*.

For an operating manual, the customer is the primary audience while technical experts and management personnel might be two secondary audiences.

The primary audience is typically some specific primary reader, able to understand but not yet informed about this particular subject. This might be a coworker in the same department working on a different product, or your supervisor, or the readership of a technical journal. Link the subject with a specific audience to which you will direct your ideas. This specific audience gives focus to your thoughts, and provides the main reason for writing in the first place.

However, the actual audience might not be what you expect. Consider a report to your supervisor. Part of this report may find its way into the department monthly status report, or into a justification of a request for additional equipment or funding. Although the primary audience must come first, possible secondary audiences must be considered. In fact, there might well be more than one secondary audience.

For example, consider the different types of audiences for which you, as an engineer, might write. In business and industry, there are countless possible primary and secondary audiences: superiors, peers, subordinates, and rivals; owners and stockholders, creditors, suppliers, and customers; government agencies; trade associations; competitors, current and potential; and others. It is your responsibility to consider which of these might read and interpret what you write. It is up to you to make any misinterpretation as difficult as possible.

The table below is included on the checklist. Use it to help make sure you have identified primary and secondary audiences before you begin to write.

Type	Peers	Superiors	Customers	Others
Primary Audience				
Secondary Audience 1				
Secondary Audience 2				
Secondary Audience 3				
Your Audience (by name)	Description			
Primary Audience				
Secondary Audience 1				
Secondary Audience 2				
Secondary Audience 3				

Once you have identified the primary and the most likely secondary audiences, you must analyze them and decide how best to write for them: what messages to send and what writing techniques to use. Let's defer a discussion of your intended messages until Chapter 4, and concentrate now on making sure you have correctly analyzed the audiences.

Here are audiences that are common in industry and business.

PEERS

Here is an audience with which you have much in common. As a primary audience, its members probably are in the same line of work as you; at least they are familiar with your work. You may know many of them personally. Presumably, they have an interest in what you do, because it affects some aspect of their own work. Your writing might be required reading.

In other words, your peers have a strong stake, professionally and emotionally, in what you write. This audience will be receptive to your ideas ... at least it will read what you have to say, even though some of its members might take exception to it.

As a secondary audience, it might be small and difficult to identify, but could include members of other departments with whom your peers converse.

SUPERIORS

Your primary audience often will be your immediate supervisor who might be interested in how well you performed some particular task, how well you met broad and specific objectives, whether you completed the work on time and within budget, and whether or not the data support the results. Different levels of management above you (potential secondary audiences) may be more or less interested in the details of your report; however, the real interest might be the effect your work could have on manufacturing yield, cost savings, or employee satisfaction.

Interests vary with each level of management, each change in responsibility, and each different personality and personal agenda. Remember that managers are almost sure to have less technical knowledge of your field than you do. Although well educated, the manager's primary education and experience may differ from yours considerably, so write accordingly. For example, if you are an electrical engineer writing about a new circuit design that increases the yield of a microchip manufacturing line, you should write a report that can be understood by a person trained in, perhaps, chemical engineering or physics.

CUSTOMERS

In most businesses, customers are always one possible primary audience. Although communication with customers might take many forms, all have common requirements. Whether a customer is reading a proposal for services, an instruction manual, or a reply to a complaint, considerations of competence, reliability, cost, and trust are foremost.

If a customer thinks you are being evasive because the commitments you make in the proposal are unclear, you risk disqualification. If your instruction manual is obscure, difficult, and infuriating, you risk a misapplied product, followed by a demand for return, service, and possibly a lawsuit if damage results. If your reply to a customer complaint is officious, unclear, and hostile, you risk losing a customer and spreading a bad reputation. Remember, the customer is the reason your business exists, and must be treated with respect and consideration at all times.

Someone in your company probably has data on the characteristics of your customers, such as education, which you should study if you will be writing to customers regularly.

Secondary audiences might include government agencies evaluating the safety of your product or its compliance with regulations, and consumer or trade organizations evaluating its worth. These secondary audiences can have a great effect on the acceptance of your product by the marketplace, and considerable care should be taken that the messages you send to them are ones you wish them to hear.

OTHERS

In addition to the audience types listed above, there are many possibilities for primary audiences and limitless ones for other secondary audiences. Here are some possible audiences you might consider.

Government agencies	Government officials	Citizens groups
Technical review teams	Professional societies	Lawyers and judges
Employees	Retirees	College students
The general public	Lobbyists	Potential competitors
Television viewers	Radio listeners	Advertisers

There are undoubtedly many more that could be listed. In fact, you might want to take a few minutes to list other possible audiences that apply to your particular situation. This list may come in handy when you are trying to analyze a real audience.

Other Potential Audiences for My Situation

_____

_____

_____

_____

_____

AUDIENCE COMPLEXITY

Each audience may be more complicated than appears at first. The viewpoint a person brings to reading has a significant effect on perception. This coloring of perception can lead to misunderstanding or clarity, to dispute or agreement, or to failure or success of the entire communication. So, in addition to correctly identifying the different audiences, you must try to analyze how homogeneous each audience is. This homogeneity or lack of it, called *audience complexity*, considers the number of different viewpoints within an audience.

Consider who your readers might be. Your writing might be read by a single person with a single viewpoint (*simple* audience) or with multiple viewpoints (*complex* audience); or by many people, each with the same viewpoint (*simple compound* audience), or with different viewpoints (*complex compound* audience); or by many people, some with single but differing viewpoints, and some with multiple viewpoints (*combined compound* audience).

These audience classifications appear in the table below.

Number of Readers	Single Viewpoint	Multiple Viewpoints	Audience Type
One			Simple
One			Complex
Many			Simple compound
Many			Complex compound
Many			Combined compound

Do not be intimidated by this table; it is intended to encourage you to think about the composition of your audience, rather than force you to perform some arcane analysis. It also appears on the *Audience Analysis Checklist* at the end of this chapter. When you prepare for a writing task, be sure to circle the type of audience you anticipate. Then follow the instructions on the checklist and answer the related questions. By taking some time to think about who might be reading your work, you have a better chance of correctly addressing your intended targets.

AUDIENCE FAMILIARITY

In analyzing the audience, you also must consider the expertise of the reader in the subject. Is the reader expert in the field, needing little preliminary explanation and tolerant of leaps of logic? Or is the reader well-educated and intelligent, but only incidentally versed in the subject, and so needing a longer introduction, but not needing to be led by the hand? Or is the reader unsophisticated in both the subject itself and in interpreting technical concepts, requiring a thorough explanation of the basics, and a basic vocabulary and treatment? For example, a reader could be a recognized expert in the field, but have no knowledge of the recent development which you are reporting.

It is critical that you assess as accurately as possible these characteristics of your audiences, both primary and secondary. If you do, your readers will find that you are talking directly to them, in the language and with a tone that makes them receptive to your messages. If you do not, your readers will think that you are talking either above or beneath them, and will find your writing either too simplistic or too complicated, and in either case, insulting. And your messages will be diluted or lost completely.

The table below expresses this concept graphically. Use it to gauge the level of sophistication with which you write. This table also appears on the checklist. Use it as part of your audience analysis.

	Education or comprehension	
Experience	High	Low
High	Sophisticated	Expert
Low	Undeveloped	Unsophisticated

AUDIENCE NEEDS

Each of the audiences you have identified wants one or more things from you. Some needs are specific and overt; these are *goals*. Other needs may not be so apparent, and may not be related to the technical content; these are *hidden agenda items*. Finally, each audience may have interests incidental to the reason for the communication that you may never know.

GOALS

Each person in each audience has specific reasons for reading your work. In engineering, this probably includes the need to learn new facts or new procedures. A reader of a report on an experiment, for example, reads to determine the facts of the experimental setup and apparatus, the outcome of the experiment, and the author's conclusions. The reader of a progress report reads to learn the status of an undertaking and to see what is predicted for the future. The reader of an operations manual reads to learn how to perform a certain task.

The meeting of these specific reasons (goals) is a joint responsibility of author and reader. The author must know the usual goals of readers of the type of document being written; the readers must understand the purpose of the document and not have unrealistic or inappropriate expectations. An operations manual for a pump is not designed to teach basic theory of rotating machinery and hydraulics. The author should not try to make it do so, and the reader should not expect that it will.

HIDDEN AGENDA ITEMS

While each reader has specific goals in mind, there are often other ambitions peculiar to that reader only. For example, the reader of the experimental report may be trying to improve experimental procedures in the department laboratory, or maybe improve the writing of experimental reports. The reader of a progress report may be trying to decide if there will be a need to request additional funding or to recommend termination or modification of the project. The reader of the operations manual may be interested in becoming so proficient that a raise will be forthcoming. The point is that each reader brings to the act of reading expectations of which you, the author, may well be unaware. These are *hidden agenda* items.

This unawareness does not excuse you from the responsibility of trying to anticipate them, however, and of addressing them in some way. For example, a diagram of the experimental apparatus and a section on problems encountered (perhaps in an appendix) may help the reader of the experimental report with a future test. A statement of future steps and a foreshadowing of the consequences of success or failure may help the reader of the progress report judge what actions may be required. A reference section giving sources of additional information on the theory of equipment operation may be quite helpful to the reader of a manual.

LANGUAGE COMPONENTS

You can tailor your writing to the audience by paying close attention to three basic components of language: vocabulary, sentence length and structure, and organization. By carefully choosing the proper combination of these three elements, you can improve the chance that your messages will be delivered.

VOCABULARY

The general rule is *Use those words which your audience already knows.* Simple enough, it would seem, but what if you are introducing new concepts that require new words? Then introduce the new words at the point they are needed (preferably not much before, and certainly not after), and define them then and there, using words the audience will understand. Of course, you will know which words need to be defined from your analysis of audience familiarity.

EXAMPLE

Original

> IN EVENT OF FIRE ELEVATORS SHALL NOT BE USED.
> INSTEAD USE MARKED EXIT STAIRWAYS.

Revised

> **IF THERE IS A FIRE**
> USE STAIRS MARKED BY EXIT LIGHT
> DO NOT USE ELEVATORS!
> THEY MIGHT GET STUCK

Explanation

IF THERE IS A FIRE:	(not IN EVENT OF)
USE STAIRS MARKED BY EXIT LIGHT	(what they *should* do)
DO NOT USE ELEVATORS!	(what they *should not* do)
THEY MIGHT GET STUCK	(and *why*)

SENTENCE LENGTH AND STRUCTURE Another way to consider your audience is through the length and structure of the sentences you use. In general, long sentences are difficult to follow; break them down into shorter ones. Vary the length to keep the reader from getting bored. Change the structure for the same reason. Endless paragraphs of "See Spot run" sentences have the same effect on a reading audience as does a monotone voice on a listening one. Of course, temper this advice with information from the audience familiarity analysis.

Original

The experiment has been completed. The results were as expected. The chemicals decomposed. The new components may be useful. They could be used for plastics or adhesives.

Revised

The experiment produced the expected results. The new components produced by the decomposition may be useful in the manufacture of plastics or adhesives.

ORGANIZATION

Organize the material in a way that makes sense and lets your reader gradually get comfortable with the subject. Introduce familiar and simple ideas first, and then proceed to the new and complex. The speed with which you proceed safely from the simple to the complex, from the old to the new, should be influenced by the results of the audience familiarity analysis.

Original

Soldering of the components into the circuit board must be done with an iron of 10 watts or less. First, insert the individual components into the proper sockets, after making sure the leads are straight. Make sure that the components are marked "J100" to indicate the high-speed version.

Revised

Select only components marked *J100*, indicating high-speed. Straighten any bent leads and insert each component into its proper socket. Solder with an iron rated at a maximum of 10 watts.

CASE STUDY

You are writing a manual for users of your company's newest product, a slide viewer. Figure 1 has been provided by the engineering department. Here is the existing instruction manual for the viewer.

Loading Transparencies into the Viewing Chamber.

Having resequenced the transparencies into the desired order, and ensured that the emulsion faces the Viewing Surface, place the ordered transparencies into the Supply Feeder on the right side of the Viewer and extend the Catch Receptacle from the right side of the Viewer by grasping the lower protruding surface and exerting force in an outward direction, ensuring that the Catch Receptacle is opened to its full extension. The first transparency may be loaded by withdrawing the Transparency

Feed Slide by pulling the serrated surface to the right using a smooth motion. Pushing the Transparency Feed Slide fully to the left with a slow and smooth motion is then required to load the transparency into the Viewing Chamber. Repeat this sequence of operation for all subsequent transparencies.

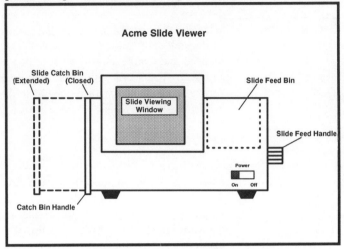

Figure 1 Slide Viewer

Is it any wonder that Uncle Joe gave up and held his vacation slides up to the dining room window?

Well, it's up to you to rewrite this instruction manual.

AUDIENCE TYPE

Now is the time to turn to the checklist for guidance. Let's consider the audience types for this operations manual.

The primary audience must be the customer who has already bought the slide viewer. However, potential customers make up an important secondary audience in two ways: first, shoppers of camera equipment often look at the manual in the store, and a poorly- or well-written manual may make the difference between a sale and a no-sale; second, buyers may have friends to whom the manual might be shown, and who may be potential customers.

Another secondary audience might be your immediate supervisor, who reviews each book before final publication and who makes judgments about you based on that review. In turn, your supervisor could be judged by the marketing manager, who cares a great deal about the effect of the instruction book on the primary and first secondary audiences ... that is, on sales.

This analysis of audience type results in the checklist being completed as shown below.

Type	Peers	Superiors	Customers	Others
Primary Audience			X	
Secondary Audience 1				X
Secondary Audience 2		X		
Secondary Audience 3		X		

Your Audience	Description
Primary Audience	actual buyers of the product
Secondary Audience 1	potential buyers of the product
Secondary Audience 2	immediate supervisor
Secondary Audience 3	marketing manager

AUDIENCE COMPLEXITY

Now that you know the audiences, analyze the complexity of each. The primary audience, the actual customer, certainly has many members, making it *compound*. It is likely that most of the members have one major viewpoint: they want to be able to operate the viewer. Although other viewpoints are possible (buying it as a gift and wanting to please or impress the recipient, for example), the need to operate the equipment successfully is the most probable. The primary audience is, therefore, *simple compound*.

The first secondary audience, the potential customer, also has many members, but also may involve many viewpoints because purchase and operation will not be the main concern of its members. Some may be shopping with the intent to purchase; some only to learn about the newest gadgets; some only to kill time; some may have no interest in a viewer until another buyer shows them the product and the book. These many members and multiple viewpoints make this audience *complex compound*.

The next secondary audience, your supervisor, obviously has only one member. But because your supervisor needs to judge you and to be judged by others based on the writing that you have done, there are multiple viewpoints. For example, you may have done a masterful job of explaining the operation of the viewer, but in a style to which the company is unaccustomed. Your supervisor may be pleased with your work but may not want to "rock the boat" by approving a departure from standard. This single member with multiple viewpoints makes this a *complex audience*.

The third secondary audience, the marketing manager, also has a single member, but has one overriding viewpoint: how will this instruction book affect sales of the viewer? Although considerations of style and protocol may be present, this one predominant concern makes this audience a *simple* one.

The resulting checklist section looks like this:

Number of Readers	Single Viewpoint	Multiple Viewpoints	Audience Type
One	2C		Simple
One		2B	Complex
Many	1		Simple compound
Many		2A	Complex compound
Many			Combined compound

AUDIENCE FAMILIARITY

Years ago, you could have counted on this primary audience, the customer, to have some technical sophistication. Then, slides were produced mainly by fairly complex manually set 35mm cameras. Now, 35mm cameras require no more technical knowledge than the old snapshot cameras. You cannot depend on the audience having even a rudimentary technical vocabulary, or ever having been exposed to any technical instruction manual.

So, the primary audience could well be relatively inexperienced in the operation of photographic equipment.

Further, because the use of slide film has increased considerably among amateur photographers, the chance of the customer not being highly educated also has increased.

This combination of low experience and low education puts the primary audience in the *unsophisticated* category of audience familiarity. These same arguments put the first secondary audience, the potential customer, in the same category.

The next secondary audience, your supervisor, should be quite familiar with the viewer, since it is a primary product of the department. Also, your supervisor is probably well-educated. This combination puts this secondary audience in the *sophisticated* category.

The third secondary audience, the marketing manager, may not be as familiar with the technical details of the viewer, being responsible for other product lines as well, but probably is well-educated. These characteristics put the marketing manager in the *undeveloped* category.

These conclusions are summarized below.

	Education or comprehension	
Experience	High	Low
High	Sophisticated	Expert
Low	Undeveloped	Unsophisticated

Your Audience	Familiarity
Primary audience	Unsophisticated
Secondary audience 1	Unsophisticated
Secondary audience 2	Sophisticated
Secondary audience 3	Undeveloped

AUDIENCE NEEDS

Let's look now at what your audience needs to get from this manual.

GOALS

All the readers of this manual want is to take a quick look at their slides before loading them into a slide tray for projecting later. They do not want or need to become technical experts on the design of the viewer.

HIDDEN AGENDA ITEMS

They may also want to show their slides to whatever unfortunate family member is closest at the time. They do not want to look incompetent. They want to work the viewer smoothly and easily, the first time. Ideally, they do not want to read the darn manual at all, but if they must they do not want to have to work at it.

OTHER NEEDS

Maybe they don't want to feel they made a mistake buying this thing; they do not want to undermine their own self-confidence. The result is shown in the following table.

Your Audience	Needs
Goals	View slides quickly and easily.
Hidden agenda items	Don't look incompetent. Have viewer operate smoothly. Don't have to read manual.
Other needs	Be reassured of buying decision. Retain self-confidence.

LANGUAGE COMPONENTS	Let's try to apply the three principles of vocabulary, sentence structure and length, and organization. Remember, the audience analysis shows that the primary audience is not necessarily well-educated nor experienced in the operation of photographic equipment.
VOCABULARY	Eliminate the references to the technical names of the parts of the viewer. Forget about Transparency Feed Slide, for example. Most people call transparencies *slides*. Can you change all references to transparency to *slide*? Get rid of words and phrases like *sequence of operation, subsequent, serrated, extension, emulsion, and ordered*. (A good thesaurus is indispensable.)
SENTENCE LENGTH AND STRUCTURE	Change all of the sentences into the same structure. In this case, make all sentences of the form "Now do this."
ORGANIZATION	Make the instructions easy to follow. Remember, the primary audience is not necessarily accustomed to following complex procedures. Collect the sentences into logical steps. For example, group all of the information about getting the slides ready to load. Then, arrange the steps into a list in the correct order.

THE RESULT

How to View Slides

in Four Simple Steps

1. Arrange your slides.

Turn each slide so that you can see the slide number.

Turn each slide so the top of the slide faces away from you.

Put the slides in a pile in the order you want to see them, with the first slide you want to see on the bottom of the pile.

2. Load your slides.

Hold the Viewer so that its window faces you.

Put the pile of slides into the bin on the right side of the Viewer so that the tops of the slides face away from you.

3. Open the catch bin.

Pull the tab that hangs below the left side of the Viewer. This will pull out the bin that will catch the slides after you have viewed them.

Make sure the bin is all the way out.

4. View your slides.

Find the small handle with ridges on the right side of the Viewer. First pull it completely out to the right, and then push it completely back in to the left. Every time you operate the handle, the next slide will appear in the viewing window.

CHECKLIST FOR AUDIENCE ANALYSIS

AUDIENCE TYPE

Type	Peers	Superiors	Customers	Others
Primary Audience				
Secondary Audience 1				
Secondary Audience 2				
Secondary Audience 3				
Your Audience	**Type**			
Primary Audience (by name)				
Secondary Audience 1				
Secondary Audience 2				
Secondary Audience 3				

AUDIENCE COMPLEXITY

Number of Readers	Single Viewpoint	Multiple Viewpoints	Audience Type
One	▓		Simple
One		▓	Complex
Many	▓		Simple compound
Many		▓	Complex compound
Many	▓	▓	Combined compound
Your Audience	**Complexity**		
Primary Audience			
Secondary Audience 1			
Secondary Audience 2			
Secondary Audience 3			

AUDIENCE VIEWPOINTS

Your Audience	Viewpoint
Primary Audience	
Secondary Audience 1	
Secondary Audience 2	
Secondary Audience 3	

Audience Familiarity

Experience	Education or comprehension	
	High	Low
High	Sophisticated	Expert
Low	Undeveloped	Unsophisticated

Your Audience	Viewpoint
Primary Audience	
Secondary Audience 1	
Secondary Audience 2	
Secondary Audience 3	

Audience Needs

Your Audience	Needs
Primary Audience	
Goals	
Hidden Agenda Items	
Other Interests	
Secondary Audience 1	
Goals	
Hidden Agenda Items	
Other Interests	
Secondary Audience 2	
Goals	
Hidden Agenda Items	
Other Interests	
Secondary Audience 3	
Goals	
Hidden Agenda Items	
Other Interests	

CHAPTER 4 — DEFINE YOUR PURPOSE

The last chapter emphasized that you should first analyze your audience - determine its makeup, its understanding of the topic, and the hidden agenda items it brings to the communication. Once you have done that, it should be relatively easy to determine how best to write in order to achieve your purpose. Unfortunately, if you do not perform the same sort of analysis on yourself and on your situation, you may *not* know exactly what that purpose is. If you don't, how can you expect your audience to know?

Although you may think you know why you are writing, you must analyze your own overt and hidden motives and agendas as thoroughly as you did those of the intended audience. This will help you to present your ideas so that you match your goals to the specific needs of the audience. When your goals and the needs of the audience coincide, communication will be most effective. You will be able to make every word, every sentence, every paragraph contribute toward effective communication.

Just as for audience analysis, there is a Checklist for Writer Analysis to ensure that you have considered all of the factors which might influence you. Neither checklist is a complete answer to all of your writing questions, but taken together they provide structure for your prewriting activities.

As in the last chapter, each part of the Checklist for Writer Analysis is explained below, and the entire checklist appears at the end of the Chapter.

WHY ARE YOU WRITING?

There are only a few basic reasons why people write. If we eliminate personal diaries, poetry, fiction, and chain letters, virtually all writing fits into one of three categories: to *inform*, to *request*, or to *persuade*. Think of the writing you do at home. The letter to Aunt Bess is to inform her what has been happening with you, her favorite relative. If she is wealthy, it may be intended to persuade her that you *are* her favorite relative. That letter complaining about the trash pickup informs the mayor of an undesirable situation and then tries to persuade the mayor to improve it. The letter to the third-grade teacher requests an appointment to review your daughter's performance. "My new address is...." "Please cancel my subscription...." "It is in the best interests of the community that you...." Inform, request, persuade. Engineers, too, write to accomplish these three objectives, even on the job.

REASONS

The table below, from Chapter 2, is repeated here because it is the basis for part of the Checklist for Writer Analysis. It lists the three categories of written communication which we will discuss, shows the different reasons for writing within each category, and lists the various document types which could satisfy each reason. These document types are not the only ones which could satisfy any given reason; they are, however, document types commonly used by engineers.

Category	Reasons	Document Types
Inform	Instruct	Manual, textbook
	Notify	Announcement, memo
	Warn	Warning, caution, letter
	Interpret	Paper, report
	Clarify	Report, memo
Request	Apply for rights	Patent application, memo, letter
	Get action	Memo, letter
	Obtain information	Memo, letter
Persuade	Make impression	Proposal, newsletter, resume
	Influence decision	Proposal, report, advertisement
	Gain position	Advertisement, proposal, memo
	Gain acceptance	Report, advertisement, paper
	Recommend	Report, memo, paper
	Sell	Advertisement, proposal

The three categories are arranged in order of increasing "rational enthalpy." That is, reasons are progressively less well defined and the document types are increasingly repeated. This shows a gradual shift of the writer's motive from mostly simple and overt to more complicated and covert. This shift roughly corresponds to the differences you encountered during audience analysis: the audience's goals, hidden agenda items, and other interests. The difference is that during audience analysis you had to guess; now you need only examine your own motives and objectives, which should be an easier task.

Let's look more closely at these three categories.

INFORM

First: *inform*. The main reason most engineers write is to inform ... to let someone else know what is happening, what is new, what is important, how to do something, or why not to do something. Most of the document types in the *inform* category are very familiar to engineers. *To inform* is the basic reason for a report, a manual, and even a newsletter. In each case, the author is in possession of facts with which the audience is either totally or partially unfamiliar.

Category	Reasons	Document Types
Inform	Instruct	Manual, textbook
	Notify	Announcement, memo
	Warn	Warning, caution, letter
	Interpret	Paper, report
	Clarify	Report, memo

It is the writer's job to tell readers what they need to know, and to do so in such a way that they realize the importance of the information. But first, the writer must determine exactly what readers need to know and also those things with which readers need not be bothered. Remember, the readers' time is limited and valuable; do not waste it with unnecessary content or detail. Get in, get to the point, and get out. A good strategy to follow is preview the material first, explain in detail next, and finally summarize the important facts.

Documents intended primarily to inform include reports and manuals, although the other documents listed above, especially textbooks, also fit the category. While this seems straightforward, consider the *caution note*. Its only objective is to inform the reader of a dangerous condition or procedure, right? Yes ... and maybe no. That may be its primary objective, but a hidden motive might be to avoid a citation for failure to post the proper signs. Or to avoid lost-time accidents. Or to avoid a union complaint, or to help defend against a lawsuit in the event of injury.

Some of your covert or hidden agenda items when writing to inform might include complying with legal requirements, complying with company policy, satisfying a personal objective, ensuring the success of a project, promoting a particular viewpoint, or avoiding problems in the future.

The most common form of writing to inform is the report. As a technical professional, it is virtually certain you will be required to write reports, possibly on a regular basis. The type of report most common in the world of engineering is the dreaded *progress report*.

During a lengthy project, you may be required to write periodic reports on how things are going. Such reports serve several purposes. They let your supervisor know if the project is progressing smoothly or if problems exist that require management attention. They let the sponsor (who is funding the project) know that the money is being well spent. They document progress so that a record will exist for reference by other engineers should the project be exceptionally successful or exceptionally unsuccessful (good engineers learn from the experiences of others). Progress reports usually take this general form:

- Project name (and/or project number, etc.)
- Date of this Progress Report (and perhaps an identifier)
- Phase or Task of the Project at the time of the Report
- Summary
- Objectives during the reporting period
- Results during the reporting period (the progress made)
- Conclusions that may be drawn from the results
- Recommendations that will affect the next Phase or Task

Although each company or organization may specify a different format, the content will be similar. The progress report tells at what stage the project is, lists what was to have been done, and describes what actually was done. If these experiences lead to conclusions and/or recommendations, they are included.

Consider the following sample progress report.

To: J. D. Whiteside, Systems Manager
Date: November 3, 1990
From: L. M. Everson
Subject: Systems Project SP-014
Building 2 Networking Project, Phase 3
Progress Report 5

Summary

During October, the Ethernet-to-Token Ring bridge requirements were defined to the satisfaction of both the Purchasing and Accounting departments. We are preparing a purchase requisition for the equipment, and expect to make initial connections near the end of November. The potentially serious problem of asbestos abatement was avoided with the help of the Facilities department which identified a diffferent cable route.

Phase 3 Objectives for October

1. Define functional specifications for Purchasing system network (Ethernet).

STATUS: Complete. Our analysis also indicates that additional network servers may be required to achieve the desired response time once the networks are connected.

2. Define functional specifications for Accounting department network (Token Ring).

STATUS: Complete

3. Finalize functional specifications of bridge between networks.

STATUS: Complete

4. Specify cable route.

STATUS: Complete. Additional cable will be needed to route around areas containing asbestos. The extra cable and installation costs will be more than offset by the avoided abatement costs.

<u>Interim Conclusions and Recommendations</u>

1. The Purchasing department network is approaching capacity; we should study other network technologies before significantly expanding the existing network.

2. The Facilities department should be requested to map locations of asbestos in Building 2.

<u>Objectives for November</u>

1. Order bridge.

2. Install and test.

3. Implement routine exchange of network traffic.

(Signed)

L. M. Everson, Project Engineer

REQUEST

Now, let's explore the category of *request*.

Category	Reasons	Document Types
Request	Apply for rights	Patent application, memo, letter
	Obtain information	Memo, letter
	Get action	Memo, letter

Engineers also write to *request*. Their requests range from the purchase of equipment to the granting of a vacation. Writing a request puts the writer at a psychological disadvantage. A request assumes that you (the writer) are at the mercy of someone (the reader) who has something you want. The problem is to ask for that something in such a way as to preserve the professional relationship between writer and reader and, of course, to get the something you want.

Writing to request becomes more difficult as the objective of the request becomes more important or more personal to the reader and the writer. The intensity of this personal interest on the part of the reader will have been determined during the audience analysis, but it is as important to determine the intensity of your own personal interest, and then to channel and control its expression.

In the table above, *personal intensity* increases as the reason for writing goes from *apply for rights* to *get action*.

Writing to *apply for rights,* for example requesting a change of address on your driver's license, is impersonal; neither you nor the clerk at the motor vehicle office has any great emotional stake in this transaction.

Writing to *obtain information* usually has a higher level of intensity. After all, you are asking someone to go to extra trouble and take time to help you. Generally, you do not know if the person to whom you are writing will be receptive or hostile, so you should write to convey both the urgency of your request and gratitude toward the person who will grant it.

Writing to *get action* has the greatest level of intensity. You probably are asking someone to go beyond the usual level of service, perhaps even to change the way in which things are done, or to do something which has not been done before. For you to have done so indicates that you have a high personal interest in the outcome; for the reader to comply may require a high personal investment in time, energy, or emotion. Sometimes, a communication may progress through all three reasons as circumstances change. Consider this example of requesting a change of address on your driver's license.

Suppose it is now the fifth time you have written. The first two times nothing happened; the next two times the change was put through with the wrong address. With each request, your emotional stake has become greater. If you began sending copies of the request to various politicians and department supervisors, the clerk's emotional stake has increased, too. During this process, your request will have changed from *apply for rights* (Please change my address), to *obtain information* (What has happened to my request?), to *get action* (I don't care what your problems are ... just get me my license!).

So, some of your hidden agenda items when writing to request might include conveying urgency, conveying gratitude, seeming helpless, threatening dire consequences, or being precise.

PERSUADE

Now for the category of *persuade.*

Writing to persuade is not as straightforward as that intended to inform or to request. In fact, persuasion may be a hidden agenda item within an informative piece. A common example of writing to persuade is a proposal, or offer for goods or services. The writer has the primary objective of persuading the audience (usually a superior or a potential customer) of the wisdom of agreeing to the proposed actions. The motive may range from something as elementary as trying to convince your supervisor that you need to have a desk near a window to something as major as trying to sell a transit system to the mayor of a major city.

Although persuasion is the primary goal, informing is usually a secondary one. The reader may need to be told reasons for agreeing. For example, the boss has to be convinced that you need to move your desk *because* you must have sunlight to combat a rare tropical disease. The mayor must be convinced to buy your transit system *because* tests show that it will be the most cost effective and the most pleasing to the ridership. In either case, you need to organize facts, structure them into logical reasons, and create a coherent argument for your case.

Some of your covert or hidden agenda items might include presenting a professional and nonthreatening image, demonstrating competence through good organization and sound language, and showing concern for quality by making sure there are no typos and by having an effectively designed document.

Writing to persuade may be made more difficult by the fact that the audience does not necessarily want to be persuaded. In fact, the audience may be hostile. These are traits you should have uncovered during the audience analysis, and now is the time to decide how to counter them. Your covert or hidden objectives might include making the audience like you, making the audience like your company, getting the audience to concede a few minor points with which you and it are generally in agreement, having the audience believe you are an expert, and having the audience feel you believe in what you are saying.

Category	Reasons	Document Types
Persuade	Make impression	Proposal, newsletter, resume
	Influence decision	Proposal, report, advertisement
	Gain position	Advertisement, proposal, memo
	Gain acceptance	Report, advertisement, paper
	Recommend	Report, memo, paper
	Sell	Advertisement, proposal

Now that you have seen that you might have covert and hidden agenda items that may or should affect your writing, use that knowledge to fill out the Checklist for Writer Analysis. As you did for the Checklist for Audience Analysis, you must identify the primary and one or more secondary categories within which you will write. Then, for each of those categories, you must identify the primary and one or more secondary reasons for writing. Then, examine your own motives and list your own covert and hidden agenda items.

Once all this is done, you can use this analysis to tailor your writing for the audience so that all of your objectives are met. In this way, you will write a successful document that does what you want and meets the needs of your audience.

CASE STUDY

Let's fill out the checklist for our slide viewer instruction manual.

CATEGORIES AND REASONS

First, determine the primary and secondary categories. This is an instruction manual, so the primary category is *to inform*.

There seems to be no request factor, as we are not asking the reader for anything. However, in line with the desire to make and hold customers, the writing should include a secondary category, *to persuade*.

Category	Reasons	Primary	Secondary 1	Secondary 2	Secondary 3
Inform	Instruct	X			
	Notify				
	Warn				
	Interpret				
	Clarify		X		
Request	Apply for rights				
	Obtain information				
	Get action				
Persuade	Make impression				X
	Influence decision			X	
	Gain position				
	Gain acceptance				
	Recommend				
	Sell		X		

Second, determine the reasons within the two categories.

Within the primary category *to inform*:

- the primary reason is *to instruct* the reader in the proper way to operate the viewer
- a secondary reason is *to clarify* for the reader the relationships between the parts of the viewer, and to

explain for the reader the proper way to arrange and turn slides for viewing.

Within the secondary category *to persuade*:

- the first secondary reason is *to sell* the viewer to the original customer by making an attractive instruction manual that can be appreciated during a shopping trip
- the second secondary reason is *to influence* others to purchase a viewer
- the third secondary reason is *to make a good impression* of the company so other products will benefit from a favorable prejudgment.

The results of this analysis are shown below and in the completed checklist at the end of this chapter.

GOALS

What, then, should the writer's goals be, given this category and reason analysis? Simply stated, the writer's primary goal is to help the buyer to operate the viewer easily. This satisfies the audience goals nicely, in that the buyer gets to see the slides and is not made to look foolish through an inability to understand the instructions. The secondary goals, then, are to make the buyer so happy with the viewer that it is freely recommended to others, and to leave with the buyer and others a favorable impression of the company. Look at the part of the checklist shown here and at the completed checklist at the end of the chapter to see these goals.

Primary Goal	Allow buyer to operate viewer easily
Secondary Goal 1	Influence buyer to recommend product to others

HIDDEN AGENDA ITEMS

Now that the primary audience has been addressed, what about the secondary audiences ... your supervisor and the Marketing Manager (remember them from the audience analysis)? These audiences are important because they may be responsible for approving the instruction manual itself, and because they have a significant influence on your career. The most important hidden agenda item, then, is to write the manual so that your supervisor approves. Note that this means *meet the deadline* as well as *do a good job*.

Next, it wouldn't hurt to establish a reputation as a good writer. Good writers are, unfortunately, rare commodities, and being one will increase your worth to your organization. Last, you know the Marketing Manager has a great deal to do with the funding of your department, the size and frequency of raises, and the probability of promotion. Because the Marketing Manager's success depends in part on how well you do your writing job, you want to make a good impression.

The Hidden Agenda Items part of the completed checklist is shown below.

| Hidden agenda item 1 | Gain supervisor's approval |
| Hidden agenda item 2 | Establish reputation as good writer |

THE COMPLETED CHECKLIST FOR WRITER ANALYSIS

The result of the analysis described in this chapter is shown below in the completed Checklist for Writer Analysis. For each writing project that you undertake, follow this or a similar procedure to make sure that you have considered all your own motives and needs. Then compare those motives to the audience analysis and make sure there is a match between your motives and needs and those of the audience. In this way, you will increase the probability of effective communication: you will get the right messages to the right audiences.

CHECKLIST FOR WRITER ANALYSIS

Name of Document	SV701-OM
Date First Draft Due	May 30, 1992
Subject	Model 701 Slide Viewer
Title	How to Use Your Acme 701 Slide Viewer
Document Type	Operating manual

CATEGORIES AND REASONS

Category	Reasons	Primary	Secondary 1	Secondary 2	Secondary 3
Inform	Instruct	X			
	Notify				
	Warn				
	Interpret				
	Clarify		X		
Request	Apply for rights				
	Obtain information				
	Get action				
Persuade	Make impression				X
	Influence decision			X	
	Gain position				
	Gain acceptance				
	Recommend				
	Sell		X		

GOALS

Primary Goal	Allow buyer to operate viewer easily
Secondary Goal 1	Influence buyer to recommend product to others
Secondary Goal 2	
Secondary Goal 3	
Secondary Goal 4	

HIDDEN AGENDA ITEMS

Hidden agenda item 1	Gain supervisor's approval
Hidden agenda item 2	Establish reputation as good writer
Hidden agenda item 3	
Hidden agenda item 4	
Hidden agenda item 5	

CHECKLIST FOR WRITER ANALYSIS

Name of Document	
Date First Draft Due	
Subject	
Title	
Document Type	

CATEGORIES AND REASONS

Category	Reasons	Primary	Secondary 1	Secondary 2	Secondary 3
Inform	Instruct				
	Notify				
	Warn				
	Interpret				
	Clarify				
Request	Apply for rights				
	Obtain information				
	Get action				
Persuade	Make impression				
	Influence decision				
	Gain position				
	Gain acceptance				
	Recommend				
	Sell				

GOALS

Primary Goal	
Secondary Goal 1	
Secondary Goal 2	
Secondary Goal 3	
Secondary Goal 4	

HIDDEN AGENDA ITEMS

Hidden agenda item 1	
Hidden agenda item 2	
Hidden agenda item 3	
Hidden agenda item 4	
Hidden agenda item 5	

Chapter 5 GET ORGANIZED

Organizing a document properly is a lot of work. Why not just "let it happen" as you write? The reason is that an organized document puts information where the reader needs it, and that is the main purpose of writing. The organization of a document affects not only the ease with which it can be understood, but also the ease with which it can be written. So, time spent organizing before you write benefits both you and your audience.

ORGANIZATION AND UNDERSTANDING

Reading is like taking a journey: understanding is the destination, and words form the highway. And as with any journey, the pleasure derived from it depends on the ease with which the route is followed. On the Interstates, uniformly designed signposts give the driver critical information about distances, road conditions, hazards, and destinations in a quickly and easily understood form.

The same logic applies to the kind of written communication found in technical areas. The more technical the subject, the more help the reader needs to understand it; actually, the more important it is to *remove roadblocks* to understanding. The key to providing help and removing roadblocks is organization.

Giving the reader the proper signs and building proper organization into a document is as much the writer's responsibility as is choosing the proper language.

ORGANIZATION AND THE AUDIENCE

The reader has a formidable task. Somewhere in all those pages, paragraphs, and words is a message which could mean success or failure. It is vital that the reader get the message easily, quickly, and accurately. In general, the reader does not have the luxury of time; there is a task to be done, and after that another, and then another.

Remember you are writing for the specific readers you identified during the audience analysis. If they are to get what you want them to get, you must make it as easy as possible. Make your writing easy to follow, provide helpful clues to its meaning and organization, and enable the reader to grasp its concepts and conclusions.

ORGANIZING AND THE AUTHOR

A well-organized document requires the author to put forth effort even before writing begins, but the reward is that the actual writing will be much easier. Organizing consists of well-defined tasks that can be handled one at a time in a logical sequence, and you - as an engineer - may find that organizing is actually less stressful than writing. It is just another analysis, like many others that you undertake frequently. So use the organizing activity to get you started on those writing jobs that you have been dreading

APPROACHES TO ORGANIZATION

Let's look at a few ways to organize oneself and one's writing. Not all apply to every writing job, but the important thing is to do *something*. Here are some suggestions.

CLUMPING

Getting organized is actually the first step in writing the first draft of your document. Gather all your thoughts, the fully-developed as well as the half-baked, and then group them into logical collections. The not-so-technical term for this process is *clumping*. Clumping related thoughts can help you to make sense out of what may seem to be an overwhelming amount of fact, data, opinion, and needs.

Let's consider an example. Suppose you are to write a proposal to help design a new product. What selling points come to mind? Here's a list:

1. experience with other products
2. similarity of the new product to existing ones
3. your expertise in managing a large project
4. your ability to complete a project within schedule and budget
5. lists of satisfied customers
6. lists of skills of employees
7. your low price
8. your leadership in the state-of-the-art
9. credentials of your employees
10. your credentials
11. the length of time your company has been around
12. the fact that your products have always outperformed the competition
13. your new, up-to-date manufacturing facilities
14. your company's good financial condition
15. your compliance with all EEO regulations

Your list might be different, longer, shorter, depending on how good you are at brainstorming ideas.

Now, what clumps can we make?

a. experience in designing products
b. expertise in managing projects
c. price
d. reputation
e. capabilities and facilities

Instead of 15 items, there are only 5. How you should organize these clumps depends on the relationship between your company and the customer. If your company is familiar to the customer, this might mean an organization of:

(a, b)

(c)

(e, d)

However, if the customer is not familiar with your company, perhaps some background information would be good at the start:

(e, d)

(a, b)

(c)

If you know that the customer is shopping price, you would emphasize that first:

(c)

(a, b)

(d, e)

Notice that there have emerged from the 15 items only 3 clumps: *(a, b); (e, d);* and *(c)*. These could be called *Experience, Capabilities,* and *Price*. Throw in an *Introduction* and a *Conclusion*, and maybe an *Appendix* or two to handle all the cost and schedule details, and you've got an outline:

Introduction

Pertinent Experience

Capabilities

Price

Appendix 1: Proposed Schedule

Appendix 2: Cost Breakdown

Now that the job has been broken down into just a few manageable tasks, work on one at a time, starting with the easiest one first. You can concentrate on quickly creating a base of writing on which you can build the more difficult sections.

TELL 'EM THREE TIMES

Was there one teacher you had whose lectures were easier to understand and remember than others? If so, that teacher probably used one of the oldest and most time-tested techniques of oral presentation ... and one which we will apply to written communication, as well. This technique is the *Tell 'Em What You're Gonna Tell 'Em; Tell 'Em; Tell 'Em What You Told 'Em* technique. Nobody knows for sure who invented it, or if anybody took credit for the name, but it works. Let's look at the three parts and how each relates to written communication.

TELL 'EM WHAT YOU'RE GONNA TELL 'EM

To a writer, this is warming up the audience. Into this category fall the preface, the foreword, the introduction, and maybe the first chapter or section. An effective writer will tell the reader in advance the topic, its importance, and perhaps conclusions, if appropriate. Also, this is a good time to explain the organization of the piece if it is particularly long or if it does not fit a standard format.

TELL 'EM

This category contains all the factual information that the author needs to pass along to the reader. For a report, it is the details of the project, the planning, the progress, the outcome, and the conclusions. For a manual, it is the step procedures, the cautions and warnings, and the illustrations that will enable the reader to accomplish the task at hand. For a proposal, it is a listing of the requirements and the responses to them, along with the marketing reasons why the company should be the chosen bidder. In short, this is the most work and the place where the rules and principles of organization that we have been discussing really pay off.

TELL 'EM WHAT YOU TOLD 'EM

This category may be the most difficult to write well. Here the writer reiterates the conclusions, reinforces the recommendations, or reviews the major steps. Remember, many people "cut right to the chase" and turn to the last page or section to get the information they need, and some people will never read any more of your document. So remember the three categories to include in any writing: *Tell 'Em What You're Gonna Tell 'Em; Tell 'Em; Tell 'Em What You Told 'Em.* If you use this technique well, you will increase dramatically the effectiveness of your writing and the chances of your success as an author.

CAUSE AND EFFECT

Organize your material into *chunks* that present first an action and then its consequence, and repeat these chunks a few times for each fact you want the reader to remember. For example, consider a proposal for which experience is counted heavily. A repeating format of "We completed the State Street project with a savings of over $100,000" will reinforce that not only does your company have experience, but experience which results in tangible benefits.

ORGANIZATION OF TYPES OF DOCUMENTS

For most types of written technical communication, there are quasi-standard organizations that you can adapt for your own use. A list of what many writers have found to work well follows. You may be able to use them as is, or you may need to modify them, or you may need to start over, depending on your document, your audience, your purpose, and your patience.

REPORT

a] Abstract
b] Conclusion
c] Introduction
d] Body
e] Summary
f] Appendixes

PROPOSAL

a] Transmittal letter
b] Executive summary
c] Identification of project and bidder
  1) Required forms
d] Intention to bid
e] Certification (Bidder will obey applicable laws, etc.)
f] Exceptions (if any)
g] Statement of work (list of things that will be done)
h] Organization
i] Cost
j] Schedule

MANUAL

a] Purpose
b] Organization

c] Explanation of method of instruction
  d] Conventions
  e] Step procedures
  f] Summary
  g] Appendix
  h] Quick reference

TUTORIAL

  a] Introduction
  b] Purpose
  c] Explanation of method of instruction
  d] Organization
  e] Lessons
     i) Narrative
     ii) Progress tests
     iii) Review
     iv) Summary tests
  f] References

TECHNIQUES OF ORGANIZATION

Let's look at techniques to help you put organization into your writing. Some, like *outlining*, will be familiar; others, such as *writing inside-out* or *typography*, may be new to you.

OUTLINING

The most commonly known method of organizing is the written outline. Let's consider a few of the more common outlining schemes. Feel free to pick and choose the parts of each that you prefer.

Paragraphs are numbered in outlines so it is easy for you to keep track of where you are. There are several conventions for numbering outlines which have been developed over many years of trial and error, and most will work fine in any situation. A short description of each follows.

HARVARD

The advantages of the Harvard style include a familiar appearance and the ability to figure out where you are by the form of the identifier in use (I, a, 1, and so on). Its disadvantages include running out of variations for really deep levels of the outline. Also, the column of text under deep headings becomes very narrow, making it harder to read and taking up a lot of pages.

I.
    A.
        1.
            a.
                1)
    B.
II.

LEGAL

Legal is what is used in most legal briefs to identify the various paragraphs clearly, but it can apply to any writing. Its advantages include the ability to define an unlimited number of levels unambiguously. Its disadvantages are that it becomes cumbersome and confusing to talk about those deep levels (Please refer to 1.3.2.4.5.1.2.1.4), and a narrowing column width.

1
    1.1
        1.1.1
2

MILITARY

Military style serves much the same purpose as the Legal style, and differs only a little. See if you can spot the differences and identify its advantages and disadvantages.

1.0
    1.1
        1.1.1
2.0

BULLETS

Sometimes, numbers and letters are just in the way, and serve no useful purpose, especially in short documents, or for ones that are being planned in a hurry. An alternative is to use *bullets* and *dashes* to indicate the topics and levels. This method provides no automatic method of keeping your place in the document; you must make the headings within the outline descriptive enough to keep you on track. Incidentally, you don't need to use a *bullet* character; an * or O will work just as well.

- •
  - ■
    - -
    - -
  - ■
- •

Writing Inside-Out

Most people write words first, and then add illustrations later to support the words. Accordingly, most people organize their thoughts around words, making outlines, lists, or whatever seems most useful to them. However, there is another way to think about written communication that has less to do with words, and more with images and concepts.

This is *writing inside-out*, based on the old saying that one picture is worth a thousand words (at least when one is making an outline).

Consider that you are preparing to write a report about a long and complex analysis of a system for authorizing release of, transmitting of, and tracking revisions to documents sent to the customer.

The process is so complex that you are having a hard time explaining it. Outlining is not much help because every time you start to write about one part of the process, you think of something else that affects that part.

So *think in images*. An illustration of an automobile engine can do more to help a person understand what a camshaft is than can pages of explanatory text, even excellent text. Some things are best grasped by the mind when they are presented as one integrated whole, all at once.

To begin, think in broad terms, and draw boxes to represent the major parts of the process, or the different players, or the different documents involved. This is the visual equivalent of brainstorming, so let yourself go! Sketch these boxes or circles or whatever on a page.

Then connect them with lines and arrows showing the flow of paper or information. Label the lines with the names of the documents or activities. If the flow indicates that some specific person or department is responsible, note that on the line. If there are deadlines associated with some of these actions, say so. If some of the actions depend on others, draw a dotted line between them, with an arrow to indicate the direction of the influence. Soon you will have a very confusing-looking diagram in front of you. Do not despair ... this is only the starting point.

First Try At Visual Outline

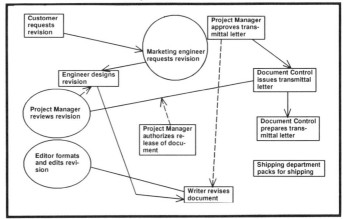

Now, step back. Could these groupings be called something specific? Redraw the diagram. How do the action flow lines and influence lines end up now? Consider changing the groupings or reexamining the actions and influences. Repeat this process until your sketches converge on a fair representation of the process.

Second Try At Visual Outline

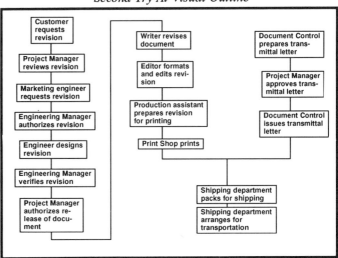

Finally, use this diagram as the map of your report. Place it up front, and relate all the following sections to it. Use this diagram to organize your writing. Use the groupings to help you name and define the various functions involved in the process, and use the action and influence lines to help you define the players.

Third Try At Visual Outline

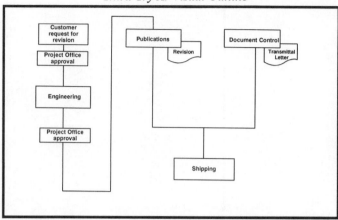

VISUAL CLUES

The page layout, choice of type style, and use of graphics influence the perception of the words, and help the reader understand the organization, which in turn helps the reader understand the messages.

PARTS OF A DOCUMENT

The easiest way to help a reader get through your document and to make sure that your messages are sent efficiently is to supply roadmaps to the document. Now let's look at these pages that are common to most documents of any appreciable length and complexity.

TITLE PAGE

The title page shows the name of the document, the name of the author, the name of the sponsoring organization, the date of publication, identifying numbers, and so on.

TABLE OF CONTENTS

A table of contents with the proper level of detail can be a great help to the reader, not only by distinguishing topics of immediate interest, but by giving a quick overview of how the document is organized, what emphasis is given to what topics (implied by the number of pages devoted to each), and what sort of items of interest might be found in the appendices.

LIST OF FIGURES AND TABLES

For the first-time reader, this provides immediate reference to visual summaries of important points. For the reader who will use the document as a reference, the lists of figures and tables will be a frequent source of specific information. Because they have the same form, only the list of figures is shown below.

Figure	Title	Page
1-1	General Arrangement of Test Facility	1-4
1-2	Test Vessel Cutaway Diagram	1-7

REFERENCES

Supplying the reader with an accurate list of sources of information is one of the most useful services a writer can render. Examples of references are shown in Chapter 6.

INDEX

The creator of an index has to balance the need for the reader to locate information against the need to keep the page count down. This balance is vital to make the index useful, because if a reader cannot find a "simple" (in the reader's mind) reference quickly, immediate frustration will result.

TYPOGRAPHY AND PAGE DESIGN

In the old days, around 1980 B.C. (Before Computers), few writers worried about typography. The draft was written out on a pad of paper, and someone typed it. But today, in the years A.D. (After Desktop-publishing), anyone with access to a personal computer can not only write, but can choose type style, page layout, and graphics.

Unfortunately, the result is often a document that looks like it was patched together from old newspaper clippings.

Fortunately, a technical person can design a pretty decent document, given the basics of typography and page design.

Let's consider some definitions.

TYPOGRAPHY

The following discussion is not a tutorial on the use of typography in a document. It is a general introduction to the vocabulary so the next time your graphic designer friend talks about serifs, leading, and Baskerville, you won't think the subject is some obscure Sherlock Holmes novel.

TYPE FAMILY

Family is at the top of the typography naming hierarchy. Family characteristics include: serif or sans-serif (with or without the little line that sits at the top and bottom of upright strokes of the letters ... that line is the serif; if it isn't there, the type is sans [without] serif); weight of the vertical and angled strokes; roundness or angularity; and many other characteristics that would surprise you by their number. There are thousands of type families in existence, many of which are distinguishable from one another only by experts.

This is Univers (sans serif).
This is Times (serif).
This is Courier (serif).
This is Helvetica (sans serif).
This is Roman (serif).

TYPE FACE

The face is a variant of the family.
This is Times.
This is Times Italic.
This is Times Bold.

TYPE SIZE

Type is specified in points. Points measure the character size and the space between the lines (see *Leading*). There are 72 points in an inch. Type is specified as *10-point* or *10pt*. A normal type size for books is 10, 11, or 12 points.

This is Times 12pt.

This is Times 8pt.

This is Times 21pt.

LEADING

Leading (pronounced ledding) is a holdover from the days of manually set type. It referred to the amount of lead (in thin blank strips) that was inserted between the lines of letters in the printing press. In other words, it means the distance between the lines of type. Type set with *zero leading* has little space between lines. Leading is specified in points, as in *2 points of lead*. More commonly, the type size and the leading are specified together. "Times Roman 10/12" (read as *Times Roman 10 on 12*) means Times Roman type, 10 points in size, set on 2 points of lead.

These sample lines are Times 10/10.
These sample lines are Times 10/10.

These sample lines are Times 10/12.
These sample lines are Times 10/12.

These sample lines are Times 10/14.
These sample lines are Times 10/14.

KERNING

Kerning means how close together the characters in a word are placed. This applies only to proportionally spaced type; for monospaced type, all characters occupy the same width. Some character pairs look better if they are close together. A and W, for example, look better pushed together a little.

PAGE DESIGN

The layout of the pages in a document makes the first impression on the reader (after the cover, of course). A page that is inviting and easy on the eye puts the reader in a much better frame of mind than a page that is dense with heavy type or crowded with small type. Here are a few guidelines to help you make pages that do their job effectively.

PAGE MARGINS

In school, you probably were told to have 1-inch margins on the left and right of the page, and 1-inch margins at the top and bottom, and that's not a bad idea. The wide side margins give your reader plenty of room to hold the page without covering up the words, and the generous top and bottom margins allow room for the page number and other identifiers. A page that has skimpy margins looks *too black* or *heavy*. On an 8-1/2-by-11-inch page, wider margins are even better. A line of type should be no longer than 6 inches and any column should be no wider than 5 or 5-1/2 inches for complicated writing. If the resulting extra margin space is added to one side, it is called a *scholar's margin*, from its usefulness in making notes.

HEADERS AND FOOTERS

The header sits in the top page margin; the footer in the bottom one. The most common footer is the page number. Other useful information that might be contained in the header or footer includes the name of the current chapter or section, the page number, the revision level, or the word DRAFT if appropriate. The idea is to give your reader a convenient place to look to find out information about the status of the document and the present location within the document.

COLUMNS

A column is a block of text on a page or data in a table. Care must be exercised in choosing the number of columns and their widths. A column that is too narrow can be as difficult to read as one that is too wide. The following examples illustrate various column layouts.

ONE-COLUMN FORMAT

Here, there is only one column, and it extends all the way across the page. A maximum width of six inches is recommended; any wider and the eye gets lost on the return trip. Such a page is useful for material which should be read slowly and carefully, like a manual or a textbook. Also, equations or other symbols can be included in lines of this width.

> Here, there is only one column, and it extends all the way across the page. A maximum width of six inches is recommended; any wider and the eye gets lost on the return trip. Such a page is useful for material which should be read slowly and carefully, like a manual or a textbook. Also, equations or other symbols can be included in lines of this width. Here, there is only one column, and it extends all the way across the page. A maximum width of six inches is recommended; any wider and the eye gets lost on the return trip. Such a page is useful for material which should be read slowly and carefully, like a manual or a textbook. Also, equations or other symbols can be included in lines of this

TWO-COLUMN FORMAT These two columns are each 2-3/4 inches wide, with a gutter (the space that separates them) of 1/2 inch. Add these up, and you will get 6 inches (there's that number again). Columns such as these contribute to ease of reading.

These two columns are each 1-3/4 inches wide, with a gutter (the space that separates them) of 1/4 inch. Columns such as these contribute to ease of reading. These two columns are each 1-3/4 inches wide, with a gutter (the space that separates them) of 1/4 inch. Columns such as these contribute to ease of reading. These two columns are each	1-3/4 inches wide, with a gutter (the space that separates them) of 1/4 inch. Columns such as these contribute to ease of reading. These two columns are each 1-3/4 inches wide, with a gutter (the space that separates them) of 1/4 inch. Columns such as these contribute to ease of reading. These two columns are each 1-3/4 inches wide, with a

POOR MULTICOLUMN FORMAT Now here's trouble. With a column width of only 3/4 inch and a gutter of 1/2 inch, the lines are so short that only a few words fit on each. The eye is in constant back-and-forth motion, fatiguing the reader. It is difficult to comprehend complex ideas because they are spread over so many lines. For technical documents, this is just a bad idea. Keep it simple.

Now here's trouble. With a column width of	only 3/4 inch and a gutter of 1/2 inch, the lines are so	short that only a few words fit on each. he eye is in	con- stant back- and- forth mo- tion, fa- tigu- ing	the read- er. It is diffi- cult to com- pre- hend	com- plex ideas be- cause they are sprea d over	so many lines. It might help to de- crease the	gut- ter, or to chang e to a small er type size, but

OTHER READER AIDS Well, you might think there can't be much more to getting organized and helping your reader through good organization. You would be almost right. Here are a few other things to think about that might be useful in certain situations.

READ ME FIRST Most technical professionals don't like to read manuals, it seems. If you are writing a manual, and some things absolutely must be understood before the user begins to wing it, maybe a simple one- or two-page insert emblazoned with

Read Me FIRST (Really!!)

would do the trick.

MAP OF BOOK   A very useful item in helping understand the organization of a document is a visual map showing the ways the various chapters and sections are presented in the book, their relations with each other, and a brief list of the key information found in them. It is a good reference tool as well as a good introductory one. Here is the map of *this* book.

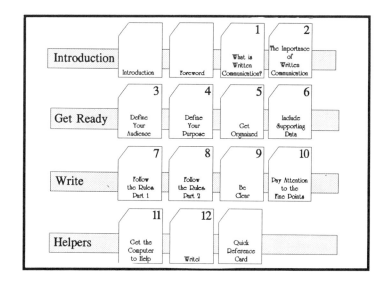

QUICK REFERENCE CARD   You probably have seen a card like the one shown on the next page if you have ever had to work on any computer system. Most programs come with one, and they are very useful if the reader will use some information in the document frequently. The quick reference card should be a size convenient for the reader to carry, and be organized so the proper information can be found in a hurry.

SUMMARY   Organizing before you write benefits everyone involved. It makes the entire communication process more efficient by contributing to clear communication between writer and reader.

Organizing makes your job as a writer easier by helping you gather and sort your thoughts and priorities so you can make sure you send your intended messages. Organizing makes your reader's task easier by grouping your thoughts into a logical structure, leading the reader through them smoothly and easily.

Among the many ways that a document can be organized, the best choice is that which communicates the intended message to the targeted audience in the most efficient way.

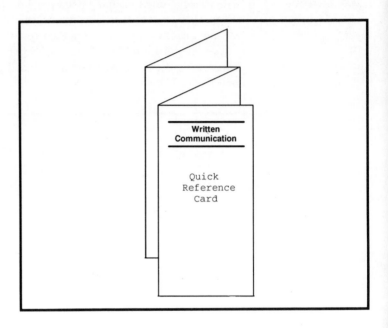

Chapter 6 — Include Supporting Data

So far, most of what we have been talking about has been the words ... the main building blocks of the document. However, most written communication in technical fields contains much more than words in paragraphs (or at least it should). Even great technical writing can be dull if it consists solely of paragraphs of text. That would be reason enough to add tables, illustrations, lists, equations, and references, but there are at least two other reasons: clarity and authority.

Clarity

A picture really is worth a thousand words, or at least it would take a lot of words to explain a concept better presented as a table or illustration. Complex ideas, mechanisms, or relationships presented visually often carry more impact than do words alone, and a visual representation may be remembered better and longer than a verbal one. Treating data visually adds to the clarity of the entire document.

Authority

Most technical writing is about data - numbers which form a set of information. The conclusions and recommendations are derived from data and the analysis of those data. Consequently, the writer should show the data in the document to establish the authority behind those recommendations and conclusions. How the data and the analysis are presented influences how the reader regards the recommendations and conclusions. Data presented concisely in an understandable and logical form foster faith in the writer's words; data presented poorly or data which do not clearly support the conclusions and recommendations engender doubt in the accuracy and completeness of the words.

Visual Elements

Data are usually presented through the use of *visual elements*. A visual element is almost anything that is not a paragraph of text. They include tables, lists, illustrations, technical drawings, equations, and references. All these forms let you use graphical and typographical techniques to enhance the presentation of data. The *typographical visual elements* of lists, tables, equations, and references put order into words and numbers; the *graphical visual elements* of illustrations and technical drawings allow you to express words and numbers in an efficient pictorial form. Add visual elements to your writing when you need to express ideas or relationships concisely in a form that will be familiar to your reader. For example, almost all of your readers will understand lists and tables, but you should avoid presenting complex equations to an audience of the general public. As in most other aspects of writing, let common sense and your audience analysis be your guide when choosing visual elements to help present your data.

TABLES

A table is a good way to show relationships among data when you need to emphasize order and relative importance among multiple variables. Consider a multicolumn table of test results, with the first column representing the test number and each remaining column representing the value of measured test variables. The reader can scan this table in either of two directions: across or down. A reader concerned with the results of a particular test can glance across the correct row and see the values for each variable. A reader concerned with a particular variable can scan down a particular column and see how the values change with the test conditions.

The table can be further divided using *spanners* that separate the rows into groups (all tests conducted at atmospheric conditions, for example), and by subdivided columns (one variable measured by two different methods, for example). The use of lines and rules within a table helps to separate the various parts, and makes it easier for the reader to grasp the concepts presented. The various ways of treating a table are shown on the next page.

LISTS

Lists are very simple types of tables, used when the organization of the data is relatively clear. All the reader needs is a place to see all of the information in one place. Lists usually do not consist of numeric data; they comprise items having the same characteristics, collected so the reader may appreciate the variety involved, or the complexity of the decisions required. Examples of lists could include all the Federal regulations that control the operation of an airport, or all the suppliers of a required part, or a selection of firms being considered as partners for a joint venture. A list may be numbered or unnumbered, depending on its purpose. Numbering implies that some order exists within the list (item 1 is more important or more desirable than item 2, for example). Unnumbered lists imply either that no such ordering exists or that the list is a set of steps to be followed in order from first to last. Examples of lists are shown on Page 6-4

Chapter 6 — Include Supporting Data

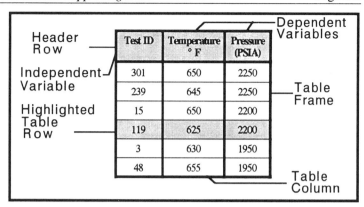

Simple Table, Row Highlighted

Simple Table, Column Highlighted

Complex Table

Numbered List, Item 1 Most Important

Hierarchy of Controlling Factors in Hiring

1. Federal regulations
2. State regulations
3. Local regulations
4. Company policies
5. General practices

Numbered List, Item 1 Most Desirable

Goals for Employee Development Program

1. Improve productivity
2. Reduce turnover
3. Develop future managers
4. Develop future senior professionals
5. Foster employee satisfaction
6. Attract new hires

Unnumbered List, Sequence Implied

How to Change a Tire

Remove jack, jack stand, and spare tire from trunk
Remove wheel covers
Loosen lug nuts
Jack up car
Place jack stands
Lower car onto jack stands
Remove lug nuts
Remove tire
Install spare
Hand tighten lug nuts
Raise car and remove jack stands
Lower car and remove jack
Tighten lug nuts
Replace wheel cover
Replace jack, jack stand, and flat tire in trunk

Unnumbered List, No Order Implied

Hit List to Complete Construction

Varnish door frames
Install walkway
Install handrails
Caulk windows
Install chimney screen

Chapter 6 *Include Supporting Data* Page 6-5

ILLUSTRATIONS

For the technical author, an *illustration* usually means a visual representation of a physical device or phenomenon, or a representation of a concept or relationship. Illustrations may be charts or graphs, drawings, or any nonverbal device used to clarify or present information. Note that this definition does not include tables. Illustrations are so varied and so widely used (and so useful) that each type deserves its own discussion.

CHARTS AND GRAPHS

Charts and graphs are visual representations of quantitative information presented in a way that can be grasped easily. Although these devices do not supply the reader with as much detail as tables (in fact, less detail sometimes is desirable), they can emphasize important points or relationships more easily, and can present data in several different ways. Also, data presented in a chart or graph may be better remembered. There are many types of charts and graphs, each suited to particular combinations of data and desired message, but all have a few common elements.

- Every chart and graph requires a title. Simple as that seems, a good descriptive title saves the reader the trouble of trying to figure out the purpose of the graphic, and it can help point out the desired message.
- At least one scale is required, since the idea is to show a relationship, even if that relationship is just *bigger*. The scale needs a descriptive title to let the reader know how the relationship is measured.
- The data presented in the chart or graph must be labeled so that the reader can identify the components of the relationship. This identification can be done by means of either a label on the component or a key that refers to some property of the component in the graphic, such as color or shading.
- If the reader will use the chart or graph as a source of quantitative information (that is, will the graph be *read* or is it just for a visual reference), then there must be a grid of some sort. A grid is especially important if the reader is expected to interpolate or extrapolate information from the data presented.
- Some charts and graphs may require a reference to the source of the data used to construct them. This reference could appear in the text, but a small note on the graphic ensures that the source information stays with the data.

Now let's look at the various types of charts and graphs, and their uses.

LINE GRAPH

The line graph is very familiar to most technical professionals. It is the most commonly used device to illustrate numerical data. It is easy to create, and presents data in a way that makes it easy for the reader to extract numerical information. The line graph is especially good at showing a continuous relationship between one or more dependent variables and one independent variable (time is frequently the independent variable). Several lines (also called curves, even if they are straight lines) on one chart illustrate the differences among continuous independent variables quite successfully.

Line Graph

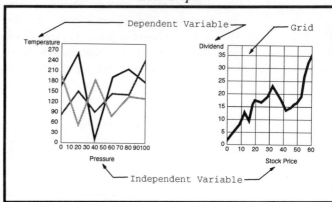

BAR CHART

Bar charts show relationships among variables at several set values of the independent variable. Bar charts may be vertical or horizontal, stacked or separated, and solid or subdivided. Each bar or set of joined bars should be separated by enough space to ensure that the reader can distinguish among them.

Bar Chart

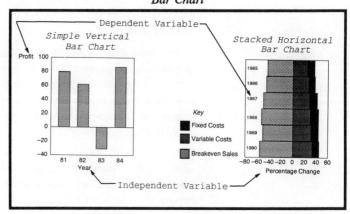

PIE CHART   When a chart must present the relative sizes of several components that make up a whole, a pie chart (also called circle chart) may be the best choice. The pie chart can be a particularly effective way to present nontechnical data or simplified technical data to a nontechnical audience. It is especially good for presenting financial information, as the circle can be crafted in the image of a coin (trite, but it works).

Pie Charts

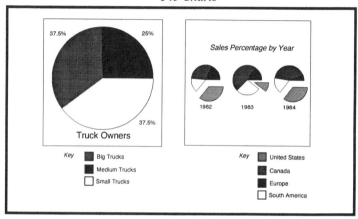

FLOW DIAGRAM   Flow diagrams usually illustrate direction. An organization chart illustrates direction of authority; a process diagram may illustrate the flow of material through a factory, or the flow of information through a computer program.

Organization Chart

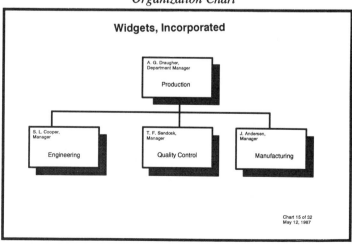

Process Diagram

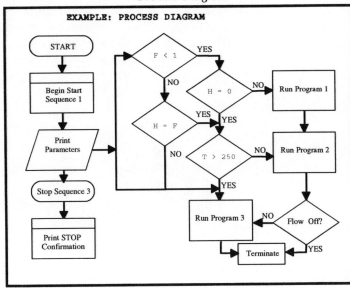

AREA CHART

An area chart combines the features of a line graph and a bar chart. As the example shows, an area chart is just a line chart showing several independent variables, with the area under each variable line shaded differently. It is this shading that allows the reader to grasp the cumulative effect of the slowly changing continuous variables, much as a stacked bar chart allows the reader to grasp the cumulative effect of a discontinuous variable.

Area Chart

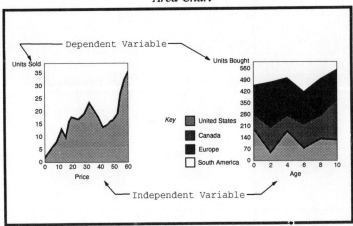

Chapter 6 *Include Supporting Data* Page 6-9

MAP CHART   There may be more to a map chart than a state map showing sales figures, as in the example below.

Map of US with Sales Figures in States

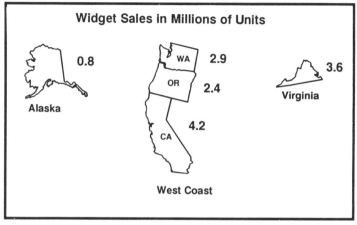

A map can be any representation of physical or theoretical space within which areas can be defined and values assigned to them. For example, the nuclear core of a power reactor is represented by a core map, with different enrichments of fuel shown in different areas.

Core Map

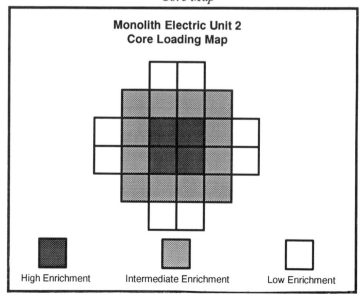

SKETCH

Here, we will call a sketch any drawing that is informal in nature, usually drawn with lines of only one weight, and with little or no dimensioning or other detail needed for manufacturing. A sketch may, in fact, be hand-drawn, but sketches are increasingly produced on a computer. Although sketches may be included in a document, it is more likely that they will be a starting point for creating either a drafting drawing or a technical illustration. In those cases where sketches appear in a document, they usually convey a feeling of informality and user-friendliness. Remember the slide viewer from Chapter 3? A sketch of the viewer might have been appropriate to illustrate that manual.

Slide Viewer Sketch

Acme Slide Viewer

Slide Catch Bin (Extended) (Closed)
Slide Feed Bin
Slide Viewing Window
Slide Feed Handle
Power On Off
Catch Bin Handle

DRAFTING DRAWING

This is a detailed and official sort of drawing that you might include in your document. If your company has a drafting department, it probably prepares all of the detailed design, assembly, and manufacturing drawings for your products. If not, perhaps you do this yourself as part of your technical duties. In any case, these drawings look like the old blueprints. They are quite detailed, have dimension lines, material specifications, notes, usually a lot of signatures of approval, and a locating grid around the periphery. Drafting drawings show so much detail that they don't usually make very good illustrations intended to clarify text. However, they are indispensable in a service or assembly manual, and may be included right in the text following their references.

In other documents, such as proposals and reports, drafting drawings usually can be relegated to an appendix, where they are available for reference, but do not interfere with the smooth flow and understanding of the text.

Drafting Drawing

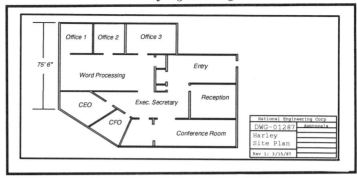

TECHNICAL ILLUSTRATION

Technical illustrations occupy a medium ground between the detailed drafting drawing and the informal sketch. A good technical illustration starts with a sketch, a drafting drawing, or a conversation. Many technical illustrations show products that have not been manufactured or even designed; these *conceptual drawings* frequently appear in proposals for major construction projects.

The purpose of a technical illustration is to clarify complex concepts or mechanisms. The technical artist does this by eliminating lines and elements that do not contribute to the message, and by enhancing those lines and elements that emphasize the message. For example, an illustration that shows how to insert Tab A into Slot B may show those two parts in heavy lines, while showing the rest in light lines. A box corresponding to the level of management currently under discussion might be highlighted by shading. Most drawings shown in manuals, proposals, and reports are of this technical illustration variety.

Technical Illustration

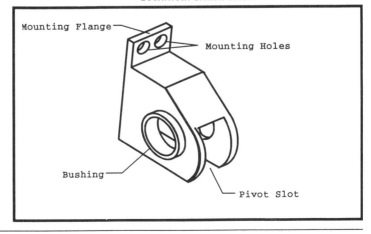

EQUATIONS

Remember this about equations: properly used, they can illuminate a subject for the proper audience better than almost anything else; improperly used, they can obscure a subject regardless of the audience. Some use of equations may be required in manuals, for example, to help the reader understand the principles of the device under discussion. Too many, though, will slow down the reader's comprehension of the process. Perhaps most of them should be relegated to an appendix. A technical report may contain some sections heavy with equations, particularly if the audience is well-versed in both the subject and the mathematics.

REFERENCES

References serve several purposes, any one of which makes the considerable effort they require worth your time. References help your readers find further information about the subject. They provide a foundation for your conclusions. They give credit where credit is due. They demonstrate a scholarly approach. And they make you look like a professional.

Here are some examples.

Reference Type	Format
Book, one author	Smith, G. Phase Diagrams. Cleveland: Acme Press, 1983.
Book, several authors	Smith, G., Havner, F. H., and Green, D. Thermal-Hydraulics. Columbus: University Press, 1986.
Journal Article	Dimling, R. K. "Writing Better Letters" Communication Digest, June 1987, 52:2, 135-7.

PART 2: WRITING

Chapter 7   Follow the Rules: Part 1

So far, we've discussed how to assemble the bones of a document. Organization, analysis of audience and objectives, selection and presentation of data - all make up the skeleton of your message. There's enough structure so the skeleton of a dog looks different from that of a human, but in either case something is missing - life! Words and how they are used determine the final form of the being created - Lassie or John Wayne, tall or short, blonde or bald, agreeable or nasty. You, the writer, are the creator. You give life to literary beings - ideas, concepts - which you hope will convince the reader of your point of view. Just as an actor who moves stiffly, speaks with an accent inappropriate to the role, or wears an inaccurate costume cannot hope to immerse the audience in the play, neither can writing with bad grammar, an inappropriate vocabulary, poor punctuation, or incorrect spelling hope to communicate its message. This chapter discusses grammar; Chapter 8 covers vocabulary, punctuation, and spelling.

GRAMMAR

Remember Miss Fischbein? She's back. Actually, she never left you, because you probably still remember the terrible time you had in English class when she tried to teach you grammar. All that diagramming of sentences, those participles that seemed to dangle all over your test paper, and objectionable (or was that objective?) clauses - I bet you still don't like to think about grammar. Even some of the best writers don't particularly like to think about it, but they have learned to live with it, and make it work for them. Grammar, after all, is merely the set of rules that allows a writer to express thoughts unambiguously, and a reader to understand those thoughts with a minimum of trouble. Remember the objective of good written communication: to communicate clearly and efficiently. Grammar, along with a few other elements such as punctuation, style, usage, and vocabulary, makes meeting that objective possible.

OK. Take a deep breath. We'll try to make this as painless as possible.

BUILDING A BRIDGE

Before we plunge right into grammar, let's consider a helpful mechanical analogy: let's build a bridge.

PARTS

A bridge consists of many parts, such as bolts, nuts, rivets, and I-beams. Each part has a set of properties, with one or more values. For example, a bolt (part) has the property of length which may assume (for our bridge) one of three values: short, medium, or long. This bolt also has the property of diameter which may assume one of two values: skinny or fat. Bolts have many other properties (tensile strength, material, threads-per-inch, left- or right-handedness, and more), but for the moment, consider only length and diameter.

FUNCTIONS

Bolts are used in many locations in our bridge, and each location requires some specific function. In order to perform that function, each property of the bolt must have a particular value. For example, a bolt which *functions* to fasten a sign to a support might have LENGTH=SHORT and DIAMETER = SKINNY. But a bolt which *functions* to connect two main girders might have LENGTH=MEDIUM and DIAMETER = FAT.

Further, regardless of the properties, the *wrong* part (a nut, for example), would be unacceptable.

There are two ways of looking at the relationship between the mechanical part and its function in the bridge.

- The *type* of part (bolt, nut ...) and the *values* of its properties (length, diameter ...) *determine* the possible *functions* it can have in the bridge.
- The *function* desired (support a sign, join girders...) *determines* which basic *type* of part is required, and what the exact *values* of all its properties must be.

MATERIAL SPECIFICATIONS

A partial materials specification for our bridge might look like this:

Part	Property	Value
Bolt	length	short
		medium
		long
	diameter	skinny
		fat

BUILDING A SENTENCE

Let's apply the bridge analogy to grammar.

PARTS

The English language also consists of *parts* (called *parts of speech*) such as nouns, pronouns, and verbs. Each of these parts of speech has a set of *properties*, each of which may have one or more *values*. For example, a *noun* (part) has the property of *number* which may assume values of *singular* or *plural*. It also has the property of *type*, which may have values of *common* or *proper*.

FUNCTIONS

Therefore, for a noun to perform the function of identifying several engineers, it would have to have NUMBER=PLURAL and TYPE=COMMON. But a noun which is to perform the function of identifying a specific state capital would have NUMBER=SINGULAR and TYPE=PROPER.

There are two ways of looking at the relationship between a part of speech and its function in a sentence.

- The *type* of part (noun, verb ...) and the *values* of its properties (number, case) *determine* the *possible functions* it can have in the sentence (subject, predicate ...).
- The desired *function in the sentence* (modify, identify ...) determines which *type* of part is required, and the exact *values* of all its properties.

PART SPECIFICATIONS

So, as for our bridge, we have a partial Part Specification for English sentences; in this case, for a *noun*.

Part	Property	Value
Noun	number	singular
		plural
	type	common
		proper

PARTS OF SPEECH

Enough of analogy. It's time to get down to the real thing: grammar.

The English language consists of these basic *parts*:

nouns, pronouns, verbs, verbals, adjectives, adverbs, conjunctions, prepositions, and *interjections*.

In order to communicate clearly, you must use these parts properly. Each of these parts has properties and allowable values; the combination of values a part has determines the function or functions it may have in a sentence. The *part specifications* for the English language are shown on the next page.

Part	Property	Value
Noun	type	common, proper, abstract, concrete, collective, count, mass
	number	singular, plural
	case	nominative, possessive
Pronoun	type	personal, relative, interrogative, demonstrative, definite, indefinite, intensive, reflexive
	number	singular, plural
	case	nominative, objective, possessive
	gender	masculine, feminine, neuter
	person	first, second, third
Verb	type	transitive, intransitive, linking, auxiliary
	principal part	infinitive, present tense, present participle, past tense, past participle
	form	progressive, intensive
	tense	present, future, past, present perfect, future perfect, past perfect
	mood	indicative, imperative, subjunctive
	number	singular, plural
	voice	active, passive
Verbal	type	infinitive, gerund, participle
Adjective	type	modifier of noun, of pronoun
		article, determiner
		predicate adjective
	comparative	positive, comparative, superlative
Adverb	type	modifier of verb, of adjective
	comparative	positive, comparative, superlative
Conjunction	type	coordinate, correlative, subordinate, adverbial
Preposition	type	single-word, multiword
Interjection		Sorry! No choices here!

NOUNS

Nouns name persons, places, things, or concepts. Nouns can be proper or common; singular, plural or collective; simple or compound; and in the nominative or possessive case.

Chapter 7 Follow the Rules: Part 1

Nouns		
Type		
Proper	refers to *specific* people, places, or things; in general, *capitalized*.	Los Angeles, Professor Smith, Harvard
Common	refers to members of large *classes* of people, places, or things; may be *concrete*, *abstract*, or *collective*; in general, *not capitalized*.	
Concrete	names *physically identified* things.	computer, boss, aardvark
Abstract	refers to an *idea* or *concept*.	hate, laziness, communication
Number		
Singular	represents only one item.	brick, Denver, dog, weather
Plural	represents more than one item.	bricks, kings, dogs, doctors
Collective	names things that come in *groups*.	flock, corps, audience
Complexity		
Simple	consists of only one word.	hat, laboratory, scientist
Compound	consists of more than one word; may be hyphenated.	high hat, castle laboratory, volt-ampere, bulls-eye
Case		
Nominative	*identifies* and is in the *nominative* case	The *laboratory* is open.
Possessive	indicates *ownership* and is in the *possessive* case.	The *laboratory's* hours are from dusk to dawn.

USES OF NOUNS Nouns perform several *functions* in sentences.

The **subject** *identifies* who or what is being discussed.	
The *tree* is a source of inspiration.	*Ambition* is the downfall of the greedy.

The **verb complement** somehow *changes* that *verb*.	
Direct Object	*Indirect Object*
The *direct object* of a verb *identifies* the person, place, or thing that has been *affected by the verb*.	An *indirect object* of a verb *identifies* the person, place, or thing toward whom or which the *action of the verb is directed*.
The frustrated chemist *booted* the *computer*.	The frustrated chemist *gave* the *computer* an ultimatum.
verb: booted direct object: computer	verb: gave indirect object: computer

The **subjective complement** explains or identifies the *subject*, and follows a *linking verb* (*is, am, are, was, were, been, will be, smell, taste*, etc.).	
Predicate Noun	*Predicate Adjective*
Stress analysis is Frieda's *obsession*.	Some say stress analysis seems *excruciating*.
subject: stress analysis	subject: stress analysis
predicate noun: obsession	predicate adjective: excruciating
connector: is	connector: seems

The **object of a preposition** *follows the preposition*. Prepositions express a *relationship* of time or space and *connect* words (*from, in, for, from, by, toward, of, at, to*, etc.).	
He sent the computer *to Antarctica*.	Antarctica looks like an indirect object because it is. An indirect object of a verb (*sent*, in this case) is an object of a preposition
preposition: to	
object of the preposition: Antarctica	

PRONOUNS

Pronouns are words used *in the place* of nouns, called the *antecedents*. Pronouns make it possible to avoid sentences like:

> The *engineer* left the *engineer's* home early and went directly to the *engineer's* office where the *engineer* could get right to work.

Instead, try:

> The *engineer* left *her* home early and went directly to *her* office where *she* could get right to work.

Pronouns come in several varieties, depending on their use in the sentence.

PERSONAL PRONOUNS

Personal pronouns are used to replace antecedents that are *identifiable personalities*. A list of personal pronouns and their properties is shown on the next page.

RELATIVE PRONOUNS

Relative pronouns *link* a dependent clause or clauses to some word in another clause (the antecedent); they *relate* the clause to the word, so the reader will understand what the dependent clause is talking about.

There are *definite* relative pronouns (*who, whose, what, which,* and *that* referring to a specific antecedent), one *indefinite* relative pronoun (*what,* referring to an unidentified antecedent), *compound* relative pronouns (any relative pronoun + *-ever,* such as *whatever, whomever* referring to a limitless number of antecedents), and *possessive* relative pronouns (*whose, of whom, of which*).

OTHER PRONOUNS

Other types of pronouns exist for different purposes, including interrogative pronouns *(who? whom? whose? which?* and *what?),* demonstrative pronouns (*this, that, these,* and *those*), indefinite pronouns (*each, any, all, few, many, neither, either, some, several, one, someone, anyone, everyone, no one* [note: two words], and *nobody*), intensive pronouns (*myself, yourself, himself, herself,* and *itself*), reflexive pronouns *(himself, myself, herself, itself),* and reciprocal pronouns *(each other, one another).*

VERBS

Verbs are the words that put *action* into writing. The verb tells *what the subject is doing* or *what is being done to the subject.* There are four types of verbs: *intransitive, transitive, linking,* and *auxiliary.* They are shown in the table on page 7-8.

Person	Case	Number	Gender	Personal Pronoun
First (the person doing the speaking)	Nominative	Singular	All	I
		Plural		we
	Possessive	Singular	All	my, mine
		Plural		our, ours
	Objective	Singular	All	me
		Plural		us
Second (the person spoken to)	Nominative	Singular	All	you
		Plural		you
	Possessive	Singular	All	your, yours
		Plural		your, yours
	Objective	Singular	All	you
		Plural		you
Third (the person, place, or thing written about)	Nominative	Singular	Masculine	he
			Feminine	she
			Neuter	it
		Plural	Masculine	they
			Feminine	they
			Neuter	they
	Possessive	Singular	Masculine	his
			Feminine	her, hers
			Neuter	its [1]
		Plural	Masculine	their, theirs
			Feminine	their, theirs
			Neuter	their, theirs
	Objective	Singular	Masculine	him
			Feminine	her
			Neuter	it
		Plural	Masculine	them
			Feminine	them
			Neuter	them

[1] There is no apostrophe; it's means it is.

Types of Verbs

Intransitive	A verb is *intransitive* if it *does not require* something for the subject to act upon (the direct object).	Frothworthy, cleaning the stain from his cummerbund, *laughed*. *Laughed* is what Frothworthy did; his laugh did not act upon anything.
Transitive	A verb is *transitive* if it *must have* something for the subject to act upon (the direct object).	Frothworthy *cleaned* the *stain* from his cummerbund. *Cleaned* is what Frothworthy did, and he did it to the *stain*.
Linking	*Linking* verbs link subjects with a word or phrase in the predicate that describes or renames the subject. They are also called *state-of-being* verbs.	
verbs of existence	*is, am, was, are, were, been, will be, shall be*	Jonathan *is* an *experienced technocrat*.
verbs of senses	*hear, smell, taste, look, see, feel, sound*	The victory yams *tasted sweet*.
verbs of status	*appear, remain, stay, become, seem, grow*	He *remained calm* in the face of all inquisitions.
Auxiliary (or helping)	*do, does, did, am, is, was, were, been, will, can, shall, could, should, would, have, has, had, may, might, must*	By working ceaselessly, Higgins *had completed* the report in record time.

TENSE

The *tense of a verb* indicates *when* the action of the verb happens. To avoid confusing the reader you must use tense consistently.

Tense of Verbs

Present	He *talks* to his laboratory glassware daily.
Past	He *talked* to his laboratory glassware daily.
Future	He *will talk* to his laboratory glassware daily.
Present perfect	He *has talked* to his laboratory glassware daily.
Past perfect	He *had talked* to his laboratory glassware daily.
Future perfect	He *will have talked* to his laboratory glassware daily.

NUMBER

The *number* of a subject - *singular* or *plural* - controls how the verb should look - singular or plural. The number of a verb is only a concern in the *present* tense; in the past tense, verbs are *independent* of number.

Number of Verbs

Present tense	Past tense
I *want* to succeed in business.	I *designed* the overpass.
She *wants* to succeed in business.	He *designed* the overpass.
Ed and Sam *want* to succeed in business.	Al and Em *designed* the overpass.

VOICE

Voice has to do with whether the subject is the *actor* (active voice) or is *acted upon* (passive voice). Normal speech usually is in the active voice. Technical writing is too often in the passive voice depriving it of life. No one is responsible, no one takes credit, no one is at fault.

Voice of Verbs	
Passive	Active
The analysis *was performed* as was requested, and the results *were transmitted* to the appropriate authorities.	Our lab *performed* the requested analysis and *sent* the results to the appropriate authorities.

There are certain acceptable reasons to use passive voice: to keep your writing interesting (all active voice makes a monotonous report), and to work around an *unknown* doer of the action.

INFINITIVES

An infinitive is the result of combining a verb with *to*, as in *to sit, to write*, and a split infinitive is what happens when the *verb* gets separated from the *to*.

Split infinitive	Infinitive
He decided <u>to</u> immediately <u>go</u> to the store.	He decided <u>to go</u> immediately to the store.

In general, it is unwise to split an infinitive, but in some instances avoiding it makes for a very awkward sentence.

ADJECTIVES AND ADVERBS

These parts of speech are like Laurel and Hardy: very different, yet frequently confused.

Adjectives and *adverbs* are words that *modify*. Think of each *ad*jective and *ad*verb as an *ad* (advertisement) for the word or words which it modifies. In our society, ads describe nearly everything, as do these *ads* in English.

These two sets of adjectives and adverbs are often *confused*.

Confusable Adjectives and Adverbs			
good (adjective)	*well (adverb)*	*bad (adjective)*	*badly (adverb)*
Incorrect Edna and Arthur operated the computer *good*.	Correct Edna and Arthur operated the computer *well*.	Correct Emil said he felt *bad* about the missed opportunity.	Incorrect Emil said he felt *badly* about the missed opportunity.
• Food *tastes good*, not well.	• You *feel well*, not good. • Things *work well*, not good.	• You *feel bad* about something, not badly.	• You *want* a Mercedes *badly*, not bad.

ADJECTIVES

Adjectives describe nouns and pronouns, and serve many purposes.

Descriptors	Answer questions like *when*, *how many*, and *how big*.	Thomas admired the *dreadful* rails. *Dreadful* tells what kind of rails.
Predicate adjectives	Follow a linking verb and describe the subject.	The spectrum analyzer was *uncalibrated*.
Articles	*The* is the *definite article*. *A* and *an* are the *indefinite articles*.	*The* identifies a specific thing (THE thing). *A* and *an* identify unspecific things (A thing or AN animal).
Determiners	Include *this*, *these*, *that*, and *these*.	*That* girder supports the entire balcony. *That* determines which girder.

Adjectives also describe the degree to which an attribute is possessed. This ability lies in the *comparative forms* of adjectives: *positive*, *comparative*, and *superlative*. Use *positive* for only one item; use *comparative* to compare two items; and use *superlative* to compare three or more items.

Change	Positive	Comparative	Superlative
Add -*er* or -*est*	The talk with Ellie was *sweet*.	The talk with Ellie was the *sweeter* of the two.	The talk with Ellie was the *sweetest* of all.
Add *more* or *less*, *most* or *least*	The talk with Alfred was *unusual*.	The talk with Alfred was *more unusual* than the one with Ellie.	The talk with Ellie was the *least unusual* I ever had.
Change of form	The dinner of "found" food was *bad*.	The dinner of frozen food was *worse*.	The dinner of junk food was *best*.

ADVERBS

Adverbs describe or elaborate on verbs, adjectives, adverbs, and entire sentences, and answer questions such as *when*, *where*, and *how* an action occurred.

The report was *overwhelmingly* interesting. (*Overwhelmingly* tells *how interesting* [a predicate adjective] it was.)	The accidental meeting with the CEO was *very* carefully arranged. (*Very* tells *how carefully* [an adverb] things were done.)
The busy supervisor read the report *quickly*. (*Quickly* tells *how* the report *was read* [a verb].	*Rapidly*, the truth became obvious. (*Rapidly* describes *how* the entire sentence came about [quickly].)

Like adjectives, adverbs can be used to indicate degrees of *comparison*.

Although some adverbs change form entirely to indicate the three forms, most are consistent in using *more* or *less*, or *most* or *least* in front of the positive form to create the other two forms.

Forms of Comparison			
Change	Positive	Comparative	Superlative
Use *more/less, most/least*.	Edmund played the piano *badly*.	Gerald played the piano *less badly* than he did the recorder.	Alfonso played the piano *most badly* of all his instruments..
Change of *form*	Edmund played the zither quite *well*.	Yet, Gerald played it *better*.	And Alfonso played it *best*.

PREPOSITIONS

A *preposition* (*to, of, with, at, to, for, over, on, contrary to, instead of, with regard to, because of,* and so on) *connects* words, and shows a relationship between them. The *object of the preposition* could be several words.

<u>Instead of</u> size, the worst problem was shape.

The old rule *never end a sentence with a preposition* is not rigid.

CONJUNCTIONS

Conjunctions connect sentences and elements of sentences, and *establish* their *logical relationships*.

Conjunctions come in four forms: *coordinate, correlative, subordinate,* and *adverbial*. The conjunction chosen and its type determine the relationship of the parts being connected.

COORDINATE CONJUNCTION

Coordinate conjunctions join parts of sentences having *equal grammatical weight* (adverbs to adverbs, for example). They coordinate the functioning of the parts as equal partners in the sentence. There are only seven: *and, but, or, nor, for, so, yet*. Each joins equals in different ways.

Coordinate Conjunction	Implication	Example
and	combination	Ann and her twin were physicists.
but	disagreement	Ann was a physicist but her twin was a painter.
or	positive choice	Ann wanted to be a physicist or a mathematician.
nor	negative choice	Ann did not like chemistry, nor did she like writing.
for	cause and effect	Ann loved physics, for she lived in a quantum state.
so	result of action	Ann loved physics, so she bought a cyclotron.
yet	contrast	Ann loved physics, yet she longed to write.

CORRELATIVE CONJUNCTION

Correlative conjunctions, too, *join* sentence elements of *equal grammatical weight*, but in a dramatic way. They travel in pairs, and are favorites of political speechwriters, for they make great persuasive language.

Correlative Conjunction	Related Coordinate Conjunction	Implies	Example
not only ... but	and	combination	Not only did Ann love physics, but she loved art as well.
both ... and	and	combination	Ann loved both physics and art.
either... or	or	choice	Ann would study either physics or art.
neither .. nor	nor	undesirable choice	Ann found that neither physics nor art interested her.
whether ... or	-none-	conditional choice	Whether Ann studied physics or art made little difference to her.

SUBORDINATE CONJUNCTION

Subordinate conjunctions act only on *independent* clauses, making one clause subordinate to the other. The resulting *dependent* clause usually *modifies* the other clause.

Subordinate Conjunction	Implication	Example
when, while, as, before, after, since, until, once, whenever	time	*Although Ann studied physics*, she took courses in art.
if, unless, even if	status	*Even if she passes physics*, she will continue to paint.
because, since	cause and effect	*Because she studied physics*, Ann felt she understood art better.
where, wherever	location	*Wherever Ann couldn't paint*, she read her physics text.
that, so that, in order that	result of action	Ann studied physics *so that she could understand the mystery of life*.
although, though, as if, even though	contrast	Ann loved art, *even though it took her away from physics*.

ADVERBIAL CONJUNCTION

Adverbial conjunctions join *independent* clauses which remain independent. The clauses are usually joined by a semicolon or even a period; however, they may be joined by commas to achieve the desired effect.

Adverbial Conjunction	Implication	Example
also, besides, furthermore, moreover	combination	Physics was Ann's first love; besides, it was fun.
then, afterward	time	Ann signed up for physics; then she added art.
otherwise	status	Luckily, the physics class was open. Ann otherwise might have become a chemist.
therefore, thus, accordingly, consequently	cause and effect	Ann was late for registration; she therefore lost the choice physics class.
however, still, nevertheless	contrast	Ann aced physics; still, she loved to paint.

INTERJECTION

Interjections express emotion. They have few restrictions on placement and punctuation.

Interjections	
Well! What happened to the pyramid analyzer?	*Well!* expresses alarm.
Good grief! I think this computer is a clone!	*Good grief!* expresses surprise.
The report has fallen into the experiment. *What now??*	*What now??* expresses distress.

SENTENCES

Now we have the basic *parts* of speech, and understand their *properties* and the possible *values* they might assume. How are these parts assembled into the sentence that best communicates to the reader?

Let's go back to the bridge. The major components (support beams ...) were made of parts with the right values of properties (high-tensile bolts and I-beams). For a bridge to carry traffic, all the major components had to be there, and put together in the right order. For example, in a suspension bridge, the cables hang from the pillars and support the roadway. Any other order of these components would result in a very odd bridge which would not do its job.

Similarly, we saw that the *parts of speech* (nouns, verbs ...) had to be made from words with the right *values* of *properties* (number, case, gender...). For language to carry communications, the right parts have to be there and put together in the right order; in this case, the resulting structure is called a *sentence*.

Further, a bridge might have a single or double roadway, and it might be a suspension bridge or a truss bridge. The *type* of bridge depends on the *classes* of its *major components*.

Sentences, too, have *classes* of their *major components*. These components and the classes of each are:

Sentence Component	Class
Subject	Simple, compound, complete
Predicate	Simple, compound, complete
Phrase	Prepositional, participial, infinitive, verb, gerund
Clause	Independent, dependent, noun, adjective, adverbial

And, like the components of a bridge, the components of a sentence must follow certain rules for the message to be transmitted efficiently and clearly.

SENTENCE STRUCTURE

The *structure* of a sentence provides the framework upon which to hang all the parts of speech. The order in which the structure is assembled and the skill with which that is done determine the message that will be sent.

There are two main sections to a sentence: the *subject* and the *predicate*. Virtually all the parts of speech eventually find their way into either the subject or the predicate.

SUBJECT

The *subject* is what the sentence is about. In most simple sentences, the subject appears within the first several words. In more complex sentences, the subject might be anywhere.

A subject may have many words, in which case it is called the *complete subject*. But the heart of every complete subject is the *simple subject*, one or two words which identify its essence.

The well-educated employee, however employed, will take the opportunity under advisement.	
Complete subject	The well-educated employee, however employed
Simple subject	employee *(the noun)*

The subject, simple or complete, can also be *compound*. A compound subject has more than one subject joined by *and, or,* or *not*, all of which are performing the same action.

The engineer and the technician investigated the mysterious malfunction.	
Compound subject	The engineer and the technician

PREDICATE

The *predicate* describes exactly what it is that the subject is involved in or further describes the subject in some way. It always contains *at least one verb*, for the verb gives action and life to the sentence. Predicates can be *complete* or *simple*; the simple predicate is simply the verb.

The engineer and the technician investigated the mysterious malfunction.	
Complete predicate	investigated the mysterious malfunction
Simple predicate:	investigated *(the verb)*

Predicates can be *compound*, consisting of more than one verb linked by *and, or,* or *nor*, all of which relate to the same subject.

The engineer and the technician investigated the mysterious malfunction and noted its cause in their notebooks.	
Compound predicate	investigated the mysterious malfunction and noted its cause in their notebooks
Simple compound predicate	investigated and noted

CLAUSE

A *clause* is a collection of words with *its own subject and predicate*. Usually, such a structure would be a sentence, but clauses occupy an interior position within genuine sentences. Clauses come in two types: *independent* and *dependent*.

Independent Clause	
Elwood activated the process computer, and poured another cup of coffee.	
first independent clause	Elwood activated the process computer
second independent clause	[Elwood] poured another cup of coffee
coordinating conjunction	and

Both independent clauses could be sentences; they are made into clauses by the coordinating conjunction, which joins them into a larger whole.

Dependent clauses depend on the rest of the sentence.

Dependent Clause	
Elwood activated the process computer while he poured another cup of coffee.	
independent clause	Elwood activated the process computer
second otherwise independent clause	he poured another cup of coffee
made dependent by the subordinating conjunction	while
entire dependent clause	while he poured another cup of coffee

Dependent clauses may act as *adjectives*, *adverbs*, or *nouns*, and may be classified as *essential* or *nonessential*. If a dependent clause can be *eliminated without altering the meaning* of the sentence, it must be <u>nonessential</u>. If eliminating a dependent clause *alters the meaning* of the sentence, it must be <u>essential</u>.

PHRASE

A *phrase* contains *neither subject nor verb*. It may function as almost any other part of speech. Because phrases consist of many words, they give the writer considerable flexibility. In each example in the table below, the phrase is in *italics*.

Absolute phrase is grammatically independent of and modifies the entire sentence.	
The computer having been made useless, the employees went about their business.	The absolute phrase tells <u>when</u> (or maybe <u>why</u>) they went about their business.
Infinitive phrase combines *to* with its *object*, and acts as an *adjective*, an *adverb*, or a *noun*.	
His goal in life was *to become chief spectrum analyzer*.	The infinitive phrase is the <u>subject complement</u>, which is a <u>noun</u>.
Continued on next page	

Continued from previous page.

Participial phrase acts as an *adjective*, and begins with either the present or past participle.	
John, *promoted recently to head assistant,* was overjoyed.	The participial phrase acts as an <u>adjective</u> modifying John.
Prepositional phrase includes any preposition, its object, and any modifiers of either. It can act as an *adverb, adjective,* or *noun.*	
The chemical plant, *with its tall towers,* is required to install warning lights.	The prepositional adverb phrase acts as an <u>adjective</u>, describing the plant.
Over the top is where the chemical division took the sales figures.	The prepositional adverb phrase acts as a <u>noun</u>, and the <u>subject</u> of the sentence.
Verb phrase is the exception: there is a verb in a verb phrase, but there is still no subject. It consists of a *verb, any auxiliary verbs* and *attendant modifiers.* It functions as a *verb.*	
Marco *had been convinced* of the safety of the car.	The verb phrase is the <u>verb</u>. Consider: Marco *doubted* the safety of the car.

SENTENCE TYPES

Sentences come in several *types,* depending on how they are constructed: *simple, compound, complex,* and *compound-complex.*

Sentence Types		
Simple	contains *only one independent clause.*	Giorgio's experiment failed.
Compound	has <u>two (or more)</u> <u>independent</u> <u>clauses.</u>; the independent clauses are joined by the coordinating conjunction *and.*	Giorgio's experiment failed, and the laboratory vanished.
Complex	has <u>one</u> <u>independent</u> <u>clause</u> <u>and</u> <u>one or more</u> <u>dependent</u> <u>clauses.</u> Note the subordinating conjunction *because.*	Because Giorgio's experiment failed, the laboratory vanished.
Compound-complex	is a <u>compound</u> <u>sentence</u> (the laboratory vanished and hairspray became available again) <u>with</u> <u>one or more</u> <u>dependent</u> <u>clauses</u> (Because Giorgio's experiment failed).	Because Giorgio's experiment failed, the laboratory vanished and hairspray became available again.

BAD WELDS AND BROKEN TOOLS

Those who build successful bridges have learned to watch for troublesome problems. Bad welds and broken tools, for instance, can cause reworking to be required and schedules to be missed.

Similarly, those who build successful sentences have learned to watch for several common errors that can cause rewriting to be required and entire meanings to be missed. See the table on the next page for several examples.

Bad Welds and Broken Tools		
Sentence fragments Do not break sentences into *fragments* and let them lie there.	Incorrect	The annual report was there. *On the desk.*
	Correct	The annual report was there *on the desk*.
Comma splices A *comma splice* tries to stuff *too much* into one sentence.	Incorrect	Brenda lit the burner, *she was ready to begin*.
	Correct	Brenda lit the burner. *She was ready to begin*.
Double negatives *Double negatives* are mathematical: two negatives referring to the same item cancel.	Incorrect	The contract *did not give* the company *no* escape from the punitive clauses.
	Correct	The contract *did not give* the company *an* escape from the punitive clauses.

CONSISTENCY

Business and technical writing must be consistent to allow the reader to follow comfortably, to identify actors and their actions and to recognize modifiers and the modified. Inconsistency leaves the reader floundering, searching out the meaning of the sentence. *Shifts* and *danglers* are two common inconsistencies.

SHIFTS

Shifts, or improper changes in *person, number, voice,* and *tense* confuse the reader.

Shift Type	Incorrect	Correct
Person	The viewpoint of the sentence changes midway through.	
	When *one* becomes a field engineer, *you* spend time outside.	When *you* become a field engineer, *you* spend time outside.
Number	The writer forgets how many players there are.	
	When *a new graduate* is hired, *they* begin in drafting.	When new *graduates* are hired, *they* begin in drafting.
Voice	Active and passive voice are mixed in the same sentence.	
	The gremlin *loved* the hydrocarbon ring, and its days *were spent* preening in a mirror.	The gremlin *loved* the hydrocarbon ring, and *spent* its days preening in a mirror.
Tense	The writer gets careless about when things happen.	
	The technician *looked* at the developing film, and then *destroys* it.	The technician *looked* at the developing film, and then *destroyed* it.
Mood	The writer is not necessarily moody, but not vigilant.	
	It is vital that the experiment *be completed* and *proves* the theory.	It is vital that the experiment *be completed* and *prove* the theory.

DANGLERS

Dangling participles, participial phrases, gerund phrases, infinitive phrases, and *elliptical clauses* are all modifiers of one sort or another which have been separated from that which they modify. To avoid danglers, make sure the thing being modified is really in the sentence and put the modifier as close to it as possible.

Dangling participle		
Incorrect	*Running*, the finish line seemed to Zeb to be light-years away.	This sounds like the finish line is running, not Zeb.
Correct	*Running*, Zeb thought the finish line seemed to be light-years away.	Now Zeb is doing the running; we managed to keep the line still.
Dangling participial phrase		
Incorrect	*Running at top speed*, the finish line seemed to Zeb to be light-years away.	Still sounds like the line is running.
Correct	*Running at top speed*, Zeb thought the finish line seemed to be light-years away.	Zeb is back in the race; the solution was the same.

CHAPTER 8   FOLLOW THE RULES: PART 2

"Sticks and stones may break my bones, but words will never hurt me!"

Remember that schoolyard taunt? As most of us know, it just wasn't true. Words *did* hurt ... a lot, sometimes. And they still can. Words, or rather the wrong words or the right words put together wrong, can hurt ... a lot. They can hurt you professionally, personally, and financially. Words can make you seem wise beyond reality, or dense beyond belief. There is an old saying, "Better to remain silent and be thought a fool, than to speak and remove all doubt." The same thing holds for writing, especially in the business and technical worlds, where words committed to paper (or electronics!) can come back to influence your career and the success or failure of your company. This sounds like powerful stuff; it is. Words are so important that entire branches of law are devoted to their use and misuse: copyright, trademark, slander, and libel. Remember that in today's increasingly litigious world, writers (especially of technical manuals) are being held responsible for injuries because of poorly written instructions. So the right word *is* important.

VOCABULARY

A vocabulary is the collection of words and meanings which a speaker, writer, or group possesses in common. It is the medium of exchange through which information is transmitted. Engineers have one vocabulary, history majors another, and street gangs yet another. A good vocabulary is not absolute; the vocabulary you need depends on your audience, although an educated person should possess a vocabulary adequate to convey intelligent, timely, and useful information in most ordinary situations.

SIMPLICITY

Simple words are preferred over jargon or unnecessarily technical words. Jargon has become a dirty word, but it has its place. Every speciality, every field of endeavor, every lifestyle comes with its own jargon, and because the group understands the jargon, there is no problem. In fact, properly used jargon actually enhances communication. Think how much talking would be needed if electrical engineers couldn't *run a smoke-test* but had to *apply power prior to systematic system testing*, or physicians had to write *take as needed* on a prescription instead of *PRN*.

If you are sure your audience will understand the jargon you use, there is no real harm if you keep it to a minimum. However, if there is *any* doubt, do not use jargon at all; use conventional language. To the reader uninitiated in the mysteries of a particular discipline, a jargon-filled document is as incomprehensible as if it were written in a foreign language - which, in a sense, it is.

PRECISION

English is a such a rich language that anyone can use the right words to say the wrong thing, or the wrong words to try to say the right thing. When either happens, the reader loses. Do everything to make sure the words you use are exactly the ones you need, with exactly the right shade of meaning to convey your message.

SHADES OF MEANING

Many words have *almost* the same meaning. Even synonyms (words that mean the same) have shades of meaning, developed through use and customary context. Writers must be aware of these subtle distinctions and use words that produce the desired effect.

Shade of Meaning	motor	engine
An electric drill gets its rotation from a motor, not from an engine. What probably drives your automobile is an engine, using either diesel or gasoline fuel.	A *motor* is any device that converts energy into mechanical motion.	An *engine* is any device that converts energy into mechanical motion and that requires fuel.

Shade of Meaning	flammable	inflammable
An example of different shades of meaning is that someone might have a flammable personality, but not an inflammable one.	*Flammable* is preferred in technical and business writing because it clearly means "easily ignitable."	Inflammable might be confused with unflammable (a nonword, but ...) or nonflammable, so be conservative and use the safer choice.

CONFUSABLE WORDS

Many words are so close in spelling or sound that they are easily used incorrectly. Some confusable words follow. Look up their meanings when you intend to use them.

affect	*Verb*: to act upon, or to influence.	The poorly calibrated instrument *affected* the readings.
effect	*Verb*: to cause *Noun*: the result.	The *effect* of the poorly calibrated instrument was an invalid conclusion.
continuous	*Adjective*: unceasing or uninterrupted.	*Continuous* test firing of the rockets led to failure.
continual	*Adjective*: steadily repeating.	*Continual* failures of the rockets led to project cancellation.
principle	*Noun*: scientific, moral, or ethical basis.	Seatbelt lockup mechanisms demonstrate the *principle* of conservation of momentum.
principal	*Noun*: someone important *Adjective*: the most important of several items	The *principal* investigator wrote, edited, and approved the report.

ECONOMY

While it is sometimes hard to start writing, it can be nearly as hard to stop. Nervousness may make you blabber when speaking; it also can make you blabber when writing. Blabber can take many forms: verbosity (too many words); pomposity (overblown words); repetition (the same thing said too many times); and written Ah-ing (too many small words, like "of" or "that") to fill in the spaces in your thinking. These habits interfere with communication. They hide messages, tire readers, and waste time. Here are a few examples.

VERBOSITY

When people talk too much, listeners get bored, tune out, start thinking of escape. The same thing can happen with writers, especially if they are uncomfortable with writing. This may sound like a contradiction, but it makes sense. If a writer has to write something impressive but hasn't done a good job of analyzing or organizing, using a lot of words seems like a good way out. Unfortunately, most readers are too busy to put up with babbling, and tune out. Don't lose your reader by wearing out your welcome, because you won't be welcome next time the reader picks up something you have blah blah blah blah blah blah blah.

Here are several ways to avoid verbosity.

1. Analyze carefully

If you have done a good job of analyzing purpose and audience, you will tend to stick to the point, and avoid drifting into extraneous areas or spending time and words on things that don't matter.

2. Organize

If you take the time to organize before you write, your writing will go much easier and quicker, and you will write about only what is important to your purpose. An outline will keep you focused and on track.

3. Read your own writing

And do it critically. If you can, set your writing aside for a few hours or days. Then read it as if you were seeing it for the first time. If you get bored and long to put it down, so will your reader.

4. Rewrite

This is the critical part of writing well. Most good writers spend more time *rewriting* than writing, cutting out the fat, reaffirming the structure, and hammering home the messages. Rewriting separates mediocre writing from excellent writing. Take the time; it is always worth the effort. Here are some examples of and remedies for verbose writing.

Verbose	Don't use too many words if at all possible.	One reason to purchase this computer is that it is faster.
Better	Use fewer words.	Purchase this computer because it is faster.

POMPOSITY

We who follow the calling of providing expert advice to the less erudite of technical professionals strive to excel in our aspirations of educational and communication excellence. Uh-huh....

We all know people who like to exaggerate their own importance by inflating their language. They never *plan a trip*; they *establish their itinerary*. They don't *go to the movies*; they *attend the cinema*. Their jobs *involve detailed analysis of operating financial statistics* (they are *accountants*).

Writers, especially in technical fields, fall into the same trap. Most of us are excited about what we do, enjoy it, and want others to share that enthusiasm. But it may be difficult to make stress analysis sound glamorous, even to ourselves. So sometimes we let a little pomposity creep into our writing to pump up our (or our company's) importance.

Pompous	Determine the relative importance of the alternative issues.	The data collected from the test apparatus upheld the initial hypothesis.
Better	Make a choice.	Test results confirmed expectations.

Chapter 8 *Follow the Rules: Part 2* Page 8-5

REPETITION

Try not to repeat words unless you intend to repeat those words repeatedly. Do not interpret consistent use of terminology as repetition, however. If you start calling an item one name, continue to use that name throughout any technical writing. In this case, using different names to avoid repetition creates confusion.

Repetitious	Try not to repeat words unless you intend to repeat those words repeatedly.	Computers can do the work of the machines which are being replaced by computers.
Better	Do not repeat words unintentionally.	Computers are replacing machines successfully.

AHHH-ING

Frequently, speakers depend on verbal crutches to get through a conversation or a presentation. In fact, we all use the same crutches occasionally. These crutches include *In point of fact*, *You know*, *Ahhh ...* , and others.

In fact, *Ahhh ...* is the classic example of *Ahhh-ing*. Speakers who don't know exactly what to say next and fear losing the listeners' attention fill in the empty spots with some sound. *Ahhh ...* does this quite nicely, but it indicates that the speaker has not formed a complete thought, or has not solidified an argument. It's a ploy for time, a stall, while the thought process catches up.

This same phenomenon occurs in written communication. Extra words sneak into writing as the writer thinks and writes at the same time. These words are not necessarily wrong in meaning or in grammar; they are just unnecessary. If you find yourself using *Ahhh...s*, reread and edit them out. Here are examples of common *Ahhh...s*.

Ahhh...s	*the, of the, that the* The status *of the* equipment at the time *of the* experiments indicated *that the* reasons for the failure could have been the improper installation of some *of the* various power connections.	*there was, there are, it is, it was* *There are* several power plants that supply the area. *There was* a failure in the electrical system.
Better	The equipment status during the experiments indicated the failure could have been caused by improper installation of various power connections.	Several power plants supply the area. A failure occurred in the electrical system.

CONTEXT

con-text (kon'tekst) n. 1. The part of a written or spoken statement in which a word or passage at issue occurs; that which leads up to and follows and often specifies the meaning of a particular expression. 2. The circumstances in which a particular event occurs; a situation. [Middle English, from Latin *contextus*, coherence, sequence of words, from the past participle of *contexere*, to join together, weave: come together + texere, to join, weave, plait.] {Morris, William (ed.), *The American Heritage Dictionary of the English Language (New College Edition)*. Boston: Houghton Mifflin Company, 1982.}

Look up almost any word in a dictionary. You probably will find several meanings. All are correct; the reader knows which meaning is intended by the context in which the word is used.

Look at the definition of context above. How do you know whether this paragraph is about *definition 1* or *definition 2*? Because this is a book on written communication, not on events. You know because of the context in which the word *context* appears. The relationship to weaving is particularly fitting; in our context, *context* means the weaving of words into meaning.

Meaning changes with context, and context depends on more than the words you put on the paper. Remember audience analysis? The audience biases determine context, too. Electricians think of conduit as a piece of small-diameter pipe; spies think of it as a channel of information. Consider the story of Al's trip to the bank.

Example	Effect of Context on Meaning
Al said he was late because of a holdup at the bank.	Because we do not know the context, we don't know if service was bad at the bank or if bandits made off with the loot.
Shaking with fear, Al said he was late because of a holdup at the bank.	We now surmise that the holdup was illegal.
Shaking with indignation, Al said he was late because of a holdup at the bank.	Now we assume the holdup was due to poor and perhaps rude service.

GENDER NEUTRAL

Most communication, especially business or technical communication, has no need to refer to gender. A writer who inadvertently assigns gender to a person or to a class creates a potentially offensive situation. The most common error is referring to a person as *he* or *she*. Consider this example:

> When the president of the company gets to work, the first thing he should do is check with his secretary. She will inform him of his schedule for the day and remind him of important activities.

The writer assumes that the president is male and the secretary is female. These stereotypical role assignments may offend the reader, reducing the chances that the message will be communicated.

To eliminate bias from your writing, first recognize that it exists and that it has no place in effective communication; then *reword* sentences and paragraphs so that gender is not even mentioned. This may involve a little reorienting of your thinking, but it is worth the effort. And, after a short while, you will find this unbiased approach comes just as easily as did your old-fashioned sexist approach. Now look at our example, *de-gendered*.

> Upon arriving at work, the president of the company should first check with the lead secretary, who will list the day's schedule and important activities.

Of course, if there is a reason to indicate gender, then do so. For example, if you are writing about particular people, gender references are appropriate. For example:

> When *Ms. Andersen*, the president of the company, arrives at work, the first thing *she* should do is check with *her* secretary, Leslie Grant, who will inform *her* of *her* schedule for the day and remind *her* of important activities.

Be careful, however. First names can be trouble later on in the document. For example, is *Leslie* male or female? How would you refer to Leslie in a later paragraph: *him* or *her*? If you aren't absolutely sure, use unbiased language.

One remedy is to use the plural:

A doctor should always carry his stethoscope.
Doctors should always carry their stethoscopes.

Another is the use of who or whose:

A person who is a doctor should have a stethoscope.

The key is to *look* for sexism and bias until avoiding it becomes natural. And even then, continue to look. Lifelong habits are hard to eliminate, but you wouldn't want your first report to your new boss to offend her/him/it ....

(This book has been written in an unbiased manner - I hope.)

WORD USE

The right words have the power to convince, inform, win contracts, and gain fame. The wrong words have the power to discourage, misinform, lose contracts, and win infamy.

MISUSED WORDS Here are a few words that are often misused.

composed	comprised
Things are *composed of* other things, not comprised of them.	A series of things *comprise a whole*; the series does not compose the whole.
The evaluation team was *composed of* the manager, the lead engineer, and the technician.	The evaluation team *comprised* the manager, the lead engineer, and the technician.
among	**between**
Among refers to *three or more* items.	*Between* refers to *only two* items.
The work was to be divided *among* the *five* contractors.	The work was to be divided *between* the *two* contractors.
who	**whom**
Use *who* to stand for a noun used as a *subject*.	Use *whom* to stand for a noun used as the *object* of a verb, preposition, or infinitive.
The CEO, *who* used to be the Treasurer, was educated in England.	The CEO, *whom* I admire very much, was educated in England.

RELATED WORDS Like people, words have relatives. And like those of people, these relatives can be opposite in nature (antonyms), similar (synonyms), or sound the same but look different (homophones).

These relatives can be useful in finding exactly the right word, in not boring the reader, and in avoiding mistakes. A few related words appear here.

Antonyms	*Antonyms* are words with *opposite* meanings. Use them to *compare or contrast* one item with another.	
	We have completed all the *required* steps.	Several *optional* ones remain to be considered.
Synonyms	*Synonyms* are words with *similar* or the *same* meaning. Use them when you must *repeat* a term often.	
	We made several *attempts* at firing the rocket.	Each *trial* ended with failure of the ignition system.
Homophones	*Homophones* are words that *sound alike* or closely alike, but have *different meanings*.	
	He *passed* the mechanics laboratory, and *wondered* about the groaning he heard from within.	He *past* the mechanics laboratory, and *wandered* about the groaning he heard from within.

PUNCTUATION

There is one more category to include in our discussion of grammar: *punctuation*. Punctuation is the finishing touch that fine-tunes the sentence to maximize clarity.

The overriding rule for punctuation is that *it must make sense*. Its only purpose is to make the communication clearer; any punctuation not contributing to that goal (or worse, detracting from it) should not be used.

It is easy to become embroiled in all of the rules and exceptions to the rules of punctuation. If you find yourself mired down inside a sentence full of commas, semicolons, and parentheses, then abandon it to the wastebasket or the <DELETE> key, and begin anew.

In the following pages you will find a brief introduction to each mark of punctuation and a tabular presentation of related guidelines, examples, and interesting notes and cautions. A careful study of these tables should get you through most of the hazards of punctuation.

PERIOD

The *period* marks the *end of a sentence;* it is the strongest of the punctuation marks.

Period		
A period is the last mark in most sentences.		
End of a declarative sentence	This is a simple declarative sentence.	
End of a mildly imperative sentence	Turn in your annual expense report monthly.	A strongly imperative sentence would end with an exclamation mark: Turn in your annual expense report monthly!
End of an indirect question	The investigator asked if anyone had noticed a stranger around the melon.	
End of a request	Will you please not adjust the altimeter without the Captain's permission.	
End of a rhetorical question	Why don't we go home.	If the statement is less a question and more a suggestion or statement of fact, use a period instead of a question mark.
Abbreviations		
One-word terms generally require a period.	Inc. Corp.	*Incorporated* *Corporation*
Most one-word units of measurement do not require a period.	Hz km	*Hertz* *kilometer*
For abbreviations of multiple-word terms, periods usually are not used.	mpg SEC	*mile(s) per gallon* *Securities and Exchange Commission*
If a sentence ends in an abbreviation, an additional period is not needed.	The defective wainscotting was made at Gotcha, Inc.	
Initials		
Follow each initial of a name with a period.	A. J. Matherson D. Kindrick	
If all 3 initials are used periods are not required.	JFK FDR	
Lists and enumerations		
In general, do not put a period after items in a list unless all items are complete sentences.	Bring the following: 1. Post-hole digger 2. Spectrophotometer 3. Electric razor	Please do the following: 1. Read all instructions. 2. Assemble the required equipment in a dry place. 3. Call your observer.
*Sometimes, lists are strung together inside of a sentence. These are called **enumerations**.*		The equipment includes: 1. post-hole digger, 2. spectrophotometer, 3. electric razor.
With other punctuation	The period always goes inside a closing quotation mark.	

Chapter 8   Follow The Rules: Part 2   Page 8-11

COLON

The *colon* is stronger than the semicolon and sometimes as strong as the period, although it never completely terminates one thought and begins another. In this use, the colon alerts the reader to an imminent, unusual event. Colons also *introduce* lists, lengthy quotations or passages.

Colon		
Introduce a clause	The experiment was a success: it transmuted junk mail into motor oil.	A colon introduces a clause that explains or expands on the first part of the sentence.
Introduce a list	The dignitaries included: the King of Maldonia, the Ambassador to Brigadoon, and the Sultan of Swat.	Rules for employees are: 1. Arrive on time. 2. Bring your own hip boots.
Introduce lengthy quotation	The 1989 Contract states the following: Apprentices shall successfully complete a training course of at least six weeks. Said training course shall have been designed by Guild members as outlined in Paragraph 3, Section 5.	
Time	The train to Bedlam leaves at 12:00 midnight. The new record for the Cakewalk is 3:19:35!	
Bible verses	Matthew 5:7 (Editors and proofreaders, please note.)	
Bibliographies	New York: Huffy-Muffin Publishers, Inc.	

SEMICOLON

The *semicolon* is a *separator* of thoughts or items. It is a little stronger than a comma, yet not quite as strong as a period. It says, "Pay attention; something interesting is going to happen."

Semicolon		
Separate series	The packing list for the trip included Bob's tunics and hat; Janice's sandals and snowshoes; and Irving's abacus.	Use of a semicolon reduces the chance of confusion about the members of the series.
	The guest list included my mother, of Paramus; my sister, of Weirton; and her husband, of Danville.	Use of a semicolon clearly separates items.
Link independent clauses	We attached shielding to the battlements and restocked the moat; they attacked, nevertheless, leaving a trail of vanilla wafers behind them.	When two independent clauses appear in a sentence, a semicolon can substitute for a joining word (*and, but, for*).
	We attached shielding to the battlements, high on their walls, and restocked the moats; but, heedless of the futility of their quest, they attacked anyway.	Even when a joining word is present, a semicolon might be used to *separate the independent clauses* if one or more of the clauses is heavily punctuated.
	Experimenting with nuclear physics can be exciting; however, it can become rather expensive.	A semicolon can *join independent clauses* connected by transitional words or phrases.

COMMA

Pity the poor comma. It is probably the most abused of all punctuation. Writers throw commas around like rice at a wedding ... and sometimes readers feel like they are walking barefoot through all that rice! There are two kinds of *comma faults*: *too many* (usually in the wrong place), and *too few*.

Comma		
Separate items in a series	The engineer requested we bring shovels, picks, oranges, and a transit. We wondered if the boss would congratulate Sally, Harry, Manfred, or Elisa.	Notice that the next-to-the-last item in the list (oranges or Manfred) is followed by a comma. Some authorities insist this *final comma* never be used unless the items themselves are made of one or more parts; others insist it always be used so to eliminate all possibility of misunderstanding, which is the goal of technical communication.
Separate adjectives	They brought in a load of fresh, blue keyboards.	Separate each adjective that describes a noun by a comma.
Essential and non-essential clauses	Commas set off *nonessential clauses*. If you can delete the clause without damage, then *set it off by commas* . If not, the clause is *essential*, so *do not* use commas.	
	Engineers, *who have analytical minds*, should have no trouble writing clear reports.	The *nonessential* clause *who follow the rules in this book* can be deleted without changing the meaning of the sentence.
	Engineers *who follow the rules in this book* should have no trouble writing clear reports.	If the *essential* clause *who follow the rules in this book* were to be deleted, the meaning of the sentence would change.
Essential and non-essential phrases	These are much like essential and nonessential clauses.	
	Fred was annoyed when his dog, *Caligula*, devoured his floppy disk.	This example means that Fred will be taking his only dog (named Caligula) to the vet.
	Fred was annoyed when his dog *Caligula* devoured his floppy disk.	This example means that Fred has more than one dog, and that he will be taking the one named Caligula to the vet.
Introductory clauses	An *introductory* clause sets the stage for the rest of the sentence, but its absence would not be ungrammatical. *Set them off by commas* unless they are so short or so obvious that no misunderstanding is likely.	
	After he had polished his disk drive thoroughly, he made his way to the greenhouse.	*After dinner* he made his way to the greenhouse.
Interrogative clauses	The time machine is fueled and provisioned, *isn't it?*	

Chapter 8 — Follow The Rules: Part 2 — Page 8-13

Comma, cont'd.		
Joining words	Sentences composed of *two independent clauses* are joined by conjunctions usually preceded by a comma.	
	She came into the room slowly, and she saw a cloud of electrons wafting through the window.	The subject of each independent clause (*she*) is expressly stated, and the conjunction (*and* or *for*) is preceded by a comma.
	She came into the room slowly and saw a cloud of electrons wafting through the open window.	The subject of the second independent clause is not expressly stated, but must be inferred from the first independent clause. The conjunction (*and*) should *not* be preceded by a comma.
Transitions	*However*, the equipment operated flawlessly.	*Therefore*, the unfortunate incident can be put behind us.
Duplicate or similar words	What the problem *is, is*n't clear.	Use a comma to separate duplicated words.
Quotations	Commas *introduce* direct single-sentence *quotations*, and continue them past any intervening comments.	
	Althea, adjusting her tiara, remarked, "I wonder how that bacterium escaped from the chains."	"Well, I assure you," fumed Ramon, fussing with his favorite tentacle, "that I checked the computer lock right before tea!"
Direct address	*Master*, I have misplaced the ghouls.	Don't worry, *Edgar*, for they are insured.
Examples	*For example*, this sentence is an example.	
Degrees, titles	Raymond B. Titherworth, Ph.D.	A. Chaser, Esq.
Salutations	Dear John,	
Large figures	Attendance at the inquest was a record 11,762.	Include commas in large figures for clarity.
	The chip center is at 11787 Silicon Place. Our company was founded in 1284.	Certain expressions, like street numbers, telephone numbers, and years (up to four digits), leave out the comma.
Dates	July 4, 1776	When the entire date is given, separate the day and year with a comma.
	June 1964	When only month and year are given, do not use a comma.
Yes and no	*No*, I shall be unable to attend the science-writers convention.	*Yes*, the philosophical examination seems like a good idea.
e.g. and i.e.	Most technical courses, e.g, circuit analysis, involve higher matematics.	The most difficult freshman EE course, i.e., Circuit Analysis 101, involves higher mathematics.

DASH

The dash has become a "do-it-all" punctuation mark -- meaning it is often used incorrectly. There are TWO marks most people call dash: the true dash, − (the *em* dash) and the hyphen - (the *en* dash). Unless you use a computer that has some aspects of desktop publishing, you probably don't have access to the em dash. In this case, it is customary to replace the em dash (−) by two hyphens (--). The names come from the printing industry, where the em dash is as wide as the letter M in the typeface being used, and the en dash is as wide as the letter N. See the entry under **HYPHEN** for the proper use of the hyphen by itself. Here, dash means either an em dash or two hyphens together. There are several uses for the dash.

Dash	
Change thought	Frank really liked his new job -- in spite of the small office -- and looked forward to his day.
Summarize series	Good grammar, spelling, and punctuation − these are the basic elements of writing.
Insert series	The basic elements of writing − good grammar, spelling, and punctuation − are essential.
Insert interpolation	Ethel explained that the CPU -- central processing unit -- was too slow to handle the data.

ELLIPSIS

Ellipses are three dots that indicate that words or sentences have been left out. In engineering or business, use them within a report or memo to reference part of another work, but take care not to distort meaning. Treat the ellipses as a *three-letter word*. A space precedes and follows an ellipsis if it is not at the end of a sentence; then, a space precedes the ellipsis and a period follows it, making it seem to consist of *four* dots.

Ellipsis		
Within a sentence	The Smyth Report stated, "The results of using benzene ... increased the yield only under strictly controlled hazardous conditions, which are not recommended."	This would imply that using benzene may not be such a good idea.
At the end of a sentence	The Smyth Report stated, "The results of using benzene along with the Gandolfite increased the yield ...."	This would imply that using benzene might be a great idea.
Hesitation in speech	I wonder ... do you think he will quit?	
Intentionally unfinished sentence	"Let me tell you ..." the harried engineer began, before he realized what he was saying.	

Chapter 8 Follow The Rules: Part 2

QUOTATION MARKS There are two types: double (") and single ('). They have several uses.

Quotation Marks		
Double quotation marks	Double quotation marks are the most commonly used. Not only are they used to indicate a direct quotation, but they serve several other purposes.	
Direct quotation	The CEO said, "Our plans to manufacture plaid mayonnaise will bring us out of our difficulties."	The CEO said that "plaid mayonnaise" will save the company.
Irony	John blamed the failure of the test on "gremlins."	
Word used as a word	The legal department reminds all employees that "Leeks-not" is the proper spelling of our vegetable-sealing product.	
Unfamiliar terms	The smallest unit of matter yet discovered is the "Quark."	
BUT later	Quarks are identified by several whimsical properties, including color.	
In references	One of my favorite Startrek episodes is "The Trouble with Tribbles."	Use double quotation marks to set off the titles of short literary works, such as short stories, television episodes, or magazine articles.
With other punctuation	Periods and commas	
	She said, "I think that this will be great news." "I think," she said, "that this will be great news."	*Always* place periods and commas **inside** the quotation marks.
	Question and exclamation marks	
	Our president asked, "What can you do to improve?" Did our president say, "I want you all to improve"?	Question and exclamation marks go **outside** the quotation marks *unless* they are part of the quotation.
	Colons and semicolons	
	Arthur said, "I'm tired"; his patience had ended.	Colons and semicolons go **outside** of the quotation marks *unless* they are part of the quotation.
Single quotation marks	Single quotation marks are used to *nest* quotations. When multiple levels of nesting occur, alternate single and double marks.	
	"What he said," John insisted, "is 'I am going to throw this wave analyzer out the window!'" She said, "But the lab notebook says, 'I spoke with Watkins today. He said, "I am not sure which reagent to use." So I chose one myself.' How can that be more clear?"	

APOSTROPHE

Use the apostrophe to indicate *possessive*, *plural*, and *contractions*.

Apostrophe			
Possession			
Use 's		For singular nouns	The *engineer's* computer is down.
		For indefinite pronouns	*Someone's* computer has a virus.
		For compound nouns and expressions	The *Engineer-in-Chief's* computer has an infection. The Widget *Company's* computer has a medical condition.
		For plural nouns not ending in *s*	The *women's* computers are down.
		For singular nouns ending in s	The *truss's design* is complete.
		For nouns not ending in s, but sounding like they do	My *conscience's* voice told me to stop playing with the computer.
		For nouns that look the same in singular as in plural	The *deer's nose* was twitching. The *deer's noses* were twitching.
Use '		For plural nouns ending in *s*	All of the *engineers'* computers are misbehaving.
		For singular proper names ending in *s*	*Descartes'* philosophy was not discovered by a computer.
		For nouns that look plural, but have singular meaning	*General Motors'* engineers all have computers.
		For compound nouns that show joint possession, show possession for the *last* noun in the series.	The *scientists and engineers'* laboratory is full of computers. The *scientist and engineer's* laboratory has only one computer.
		For compound nouns that show individual possession, show possession for *every* noun in the series.	The *scientists' and engineers'* laboratories are both full of computers. The *scientist's and engineer's* laboratories each have one computer.
Plurals			
Use 's		For letters	Watch your *p's* and *q's*.
Do not use 's		For words referred to as words	No *ifs, ands* or *buts*.
		For numbers and figures	Computers appeared in the *1930s*. All of the *3s* come out as *6s*!
		For abbreviations and acronyms	The topic is the *ABCs* of life.
With many pronouns	**Correct**		**Incorrect**
	its		it's (means "it is")
	yours		your's (no such word)
	your		you're (means "you are")
	whose		who's (means "who is")
	theirs		there's (means "there is")

Apostrophe, cont'd		
Do not use *'s* with a word ending in *s* that is used as a descriptor.	The *writers guide* is available on a computer.	The *citizens committee* for better computing meets weekly.
BUT if the plural word does not end in *s*, use an apostrophe.	The *women's guide to writing* is available on line.	The *people's committee for better computing* meets weekly.
Do not create possessives for *things*.	The *truss's strength* lies in the design.	*would be better as:* The *strength of the truss* lies in the design.
Contractions	don't	do not
	isn't	is not
	o'clock	of the clock

HYPHEN Use the hyphen to *join words* into a single unit.

Hyphen		
Avoid ambiguity	The conference was on the state of the art computer technology. *could mean* The conference was on the state-of-the-art computer technology. *or* The conference was on the state of the art-computer technology.	Use a hyphen if omitting it results in ambiguity.
Compound Modifier	Two or more words that express a single thought are *compound modifiers*. The need for a hyphen depends on their position in the sentence.	
	Engineering is a *full-time* job. *But* The engineer worked full time.	Link component words of a compound modifier that comes *before a noun*.
	An *easily remembered* rule is "i before e except after c."	BUT if one word ends in -ly, a hyphen usually is not needed.
	Socio-economic status Military-industrial complex	Link *two thoughts that are combined*.
	Afro-American *but not* Latin American (a recognized heritage)	Connect *two nationalities* or *heritages*.
	shell-like (*not* shelllike)	Divide words having *three or more identical letters* together.
In numbers	fifty-seven	Use a hyphen *in spelled-out numbers* when one word ends in -y.
Suspensive hyphenation	He received a *3- to 5-day* suspension. He had a *short- and long-range* plan.	Use hyphens to express a *range of possibilities*.

PARENTHESES

Use parentheses to *insert additional information not covered in the main text or in a footnote or endnote.*

Parentheses		
Sentence fragment	If the parenthetical information is a *fragment* of a sentence *(like this)* do not capitalize the first word or end the fragment with a period.	This *(A quick example.)* is incorrect.
Complete sentence	(If the parenthetical information is a *complete sentence* that is independent of the rest of the information, treat it as a sentence.)	Capitalize the first word and end with a period.
	If the parenthetical information is a complete sentence but *depends on the surrounding text (this is one example)*, treat it as a phrase.	Do not capitalize the first word and do not end with a period.
For numbers	After thirty *(30)* days, the two *(2)* firms will agree a to partnership of either sixty *(60)* percent or forty *(40)* percent.	Use parentheses to *enclose the Arabic equivalents* of numbers in legal documents.
Listings	*(1)* Locate ON switch. *(2)* Turn ON switch clockwise until detent is reached. *(3)* Monitor operation on Meter B.	Use parentheses to *set off the numbers* of items in a list, unless you can use other typographical approaches.
With other punctuation	He came from the Pittsburgh (PA) area. *not* He came from the Pittsburgh, (PA) area.	Never *precede* an opening parenthesis with any other punctuation.
	The auditor (the treasurer's paramour) was not very diligent. *not* The auditor (the treasurer's paramour), was not very diligent.	*Follow* a closing parenthesis with other punctuation only if the sentence would have needed anyway.
	The reporter was from the Pittsburg (Kansas) TIMES. *but* The reporter was from Pittsburg, Kansas, and worked for the TIMES.	Enclose parenthetical information within a *proper name* only when needed to avoid confusion.

BRACKETS

Brackets are punctuation marks rarely used in technical writing (except in equations [showing nested terms, for example]).

Brackets		
Comments or corrections	The corporate report stated, "... the p.... of this division *[the Airframe Division]* could be increased."	
	The corporate report stated that, "... the profit of this division could be incensed *[sic]*."	
For parentheses within parentheses	The vendor recommended a vendor (Great Big Machines *[a Fortune 500 company]*) for our systems needs.	

Chapter 8 — Follow The Rules: Part 2

EXCLAMATION MARK The exclamation mark conveys relatively strong emotion and is rarely used in business or technical communication.

Exclamation Mark		
Add emphasis	The experiment failed because the computer failed!	Sounds like the writer was trying to shift blame onto the poor computer.
Exclaim	Good Grief! The computer crashed!	Or words to that effect...
Question as exclamation	Can you believe that computer wrecked my experiment!	No ...

QUESTION MARK The question mark indicates a question and draws attention to an uncertainty.

Question Mark		
Direct question	Did you crash the computer? You did crash the computer, didn't you?	The question mark indicates a *question* and draws attention to an uncertainty.
	He asked me if I had crashed the computer. Will you please not crash the computer.	However, *indirect questions* and *requests* do not get a question mark.
With quotation marks	I asked him, *"Did you crash the computer?"*	Place inside quotation marks if the quotation itself is a question.
	Was it you who said, *"I crashed the computer"*?	Place outside quotation marks if the quotation is not a question.

SPELLING Spelling seems to have an odd combination of a lot of rules and a seemingly equal number of exceptions. However, neither rules nor the exceptions are an excuse for misspelled words. That's why the first rule in spelling is: *buy a good dictionary*. And the second rule is: *use it*. The first spelling given in the dictionary is preferred. And look at the meaning to make sure you are using the right word.

"But I have a computer with a spelling checker," you protest, "Won't that take me off the hook?"

Sorry. The spelling checker (or *spell checker*) is a wonderful invention, but it is not foolproof. (See Chapter 10 for more information.) Besides, what will you do when the spelling checker presents you with four options? How will you know which one to choose? There is no avoiding it: you need to learn a few basic rules of spelling.

GENERAL RULES Following are several basic rules to help you spell correctly some (but not all) exceptions to those rules.

General Spelling Rules

i before e

Rule 1: i before e		Rule 2: except after c		Rule 3: or when the ie/ei is sounded like a	
lieu	orient	conceived	receipt	neighbor	sleigh
achieve	piece	deceit	received	inveigh	vein
friend	quotient			freight	weigh
review	retrieve			skein	
transient	yield				

Exceptions

breaks Rule 1		breaks Rule 2		breaks Rule 3
seize	weird	financier	sufficient	No exceptions
their	neither	science	conscience	
either	forfeit	deficient	efficient	
height	counterfeit			

Endings

Many languages change the endings of words to indicate case. Although English does this in several instances, many word endings seem to vary for no readily apparent reason. Memorizing is the solution.

-ible vs -able: Most words end in *-able*, but several hundred end in *-ible*.

-ible		-able		
divisible	flexible	acceptable	agreeable	available
collectible	permissible	adjustable	adaptable	probable
accessible		advisable	admirable	valuable

-efy and -ify: Only <u>four</u> words end in -efy.

-efy		-ify
liquefy	putrefy	All the rest end in -ify.
rarefy	stupefy	

-cede, -ceed, -sede: Only <u>four</u> words do *not* end in *-cede*.

-ceed		-sede (the only word)	-cede
exceed	proceed	supersede	All the rest end in -cede.
succeed			

Changing the final Y to I

Rule 1: For words ending in -y and preceded by a consonant	Example		Exception: If the *suffix* starts with i, keep the y. (-ing, -ish ...)	
1: change the y to i 2: add the suffix (-hood, -ness, -ful ...)	likely happy	likelihood happiness	try fly	trying flying

Rule 2: For words ending in -y preceded by a vowel	Example		Exception	
1: keep the y 2: add the suffix	boy play	boyhood playful	day lay	daily laid

Chapter 8 — Follow the Rules: Part 2

General Spelling Rules, cont

The silent E

Rule	Example			Exception	
Rule 1: Before a suffix that starts with a vowel (-ition, -ing, -ism, -ion, -ed, -al ...) , drop the silent e.	approve commune owe idle	approved communed owed idled	approving communing owing idling	cue dye Europe hoe	cueing dyeing European hoeing
Rule 2: Before suffixes that start with a consonant (-ment, -ness, -hood, -ful ...), keep the silent e.	crude arrange		crudeness arrangement	awe	awful
Rule 3: For words ending with a soft -ce or -ge, when adding -able or -ous, keep the silent e.	advantage service manage		advantageous serviceable manageable	judge	judgment

The final consonant

Many English words end in a consonant. In some situations, that consonant *must* be doubled before a suffix can be added; in other cases, it *must not* be doubled.

Rule	Example		
Rule 1: For *one-syllable* words that **end in one consonant**, before adding a *suffix that begins with either a vowel or y* (-ed, -ing. -ish, -y, -al ...) , double the consonant.	top drip clan	topped dripped	topping dripping clannish
Rule 2: For *more-than-one-syllable* words that **end in one consonant** following one vowel and *are accented on the last syllable*, before adding a *suffix that begins with either a vowel or y* (-ed, -ing., -ish, -y, -al ...) , double the consonant.	begin concur repel transfer	concurred repelled transferred	beginning concurring repelling transferring
Rule 3: For *one-syllable* words that *end in one consonant* following *one vowel*, before adding a suffix that **does not begin with either a vowel or y** (-ment, -ness, -hood ...), do not double the consonant.	new ship		newness shipment
Rule 4: For *more-than-one-syllable* root words that **end in one consonant** following one vowel and **are not accented on the last syllable**, before adding a suffix that **begins with either a vowel or y** (-ed, -ing. -ish, -y, -al ...) , do not double the consonant.	profit cancel boil	profited canceled boiled	profiting canceling boiling
Rule 5: For any word **ending in -x**, do not double the x.	tax	taxed	taxing
Rule 6: For words **ending in a double consonant**, do not double the consonant.	recess mill	recessed milled	recessing milling

General Spelling Rules, cont		
The final consonant, cont'd		
Rule	*Example*	
Rule 7: For words that **end in a single consonant** following **more than one vowel**, do not double the consonant.	float stout	floated floating stoutly stoutish
Rule 8: For words that **end in more than one consonant**, do not double the consonant.	sand farm	sanded sanding farmed farming

WHEN IN ROME ... A less prevalent problem than plain old misspelling is confusion between the British and American spelling of many words. While some people seem to think it is sophisticated to use the British spelling in the US, the best rule to follow is "When in Rome ...." If you are writing for a United States audience, use the American spelling.

American (U. S.) vs. British Spelling		
	American (U. S.)	**British**
When you look up a word in an American (U. S.) dictionary, you may find a secondary spelling. Usually, this is the British spelling. British spelling is inappropriate when writing for a U. S. audience.	anemia inquiry organize apologize color labor odor center theater plow traveled judgment	anaemia enquiry organise (and organisation) apologise colour labour odour centre theatre plough travelled judgement

Chapter 9 Be Clear

Now that you know the formal rules and the exceptions to them, it is time to look at techniques and style. Let's go back to that bridge again.

Suppose you are the architect. You know the types and properties of all the material needed to make a bridge. You know what the bridge is to span, and what traffic it will bear. And you know the budget and schedule within which the bridge is to be constructed. Beyond that, you have relative freedom to design the bridge as you would like. As long as the bridge does its job, not many will care if it is a truss or suspension, or steel or concrete, or classical or modern. But as a good architect, you want to ensure that the bridge not only does its job; you want to ensure that the bridge will be a thing of beauty, fitting in well with its environment, and making the sort of architectural statement that will enhance your reputation and that of the political organization paying for it.

It's the same way with sentences. You now have all the tools to build a perfectly correct English sentence. You know what the sentence needs to do; you know what words you may use safely; and you know how to string the words into valid sentences. You might even know how to spell most of them.

Now comes the fun part; you get to design. Also, now comes the bad news. Just as that bridge architect needs to understand the principles of esthetics to appreciate the visual impact the bridge will have on the community, so you need to understand and apply the principles of general communication to appreciate the impact your writing will have on the reader.

HIDDEN VERBS

Just as passive voice can hide action, so can indirectness. Look at your sentences; do they allow the verb to do its job, or is the real action buried under a pile of phrases, clauses, prepositions, and circumlocutions?

Incorrect	Correct
All the parties present were in agreement that the program should be terminated expeditiously.	Everyone agreed to end the program.
The Lead Engineer must be cognizant of all manufacturing procedures.	The Lead Engineer must know all manufacturing procedures.
Distribution of the observation forms was completed last Tuesday.	We distributed the observation forms last Tuesday.
In compliance with your directive, all computers have been irradiated.	As you directed, we have irradiated all computers.

INVENTED WORDS Those who speak English, especially Americans (and especially Americans in business or government), love to make up words to fit the occasion. And for most informal spoken communication, no harm is done. But for more formal written communication, please stick to real words. These invented words have the same drawback as jargon: not all your readers may understand them.

VERBIZING A favorite way to make new words is to *verbize* (there's one now!). The process is easy: take a word (usually a noun) and add *-ize* or, easier yet, just use it as is. Here are some to avoid.

Incorrect	Correct
Don't *verbize* nouns.	Don't *make* nouns into verbs.
She *gifted* him with flowers on Lincoln's birthday.	She *gave* him a *gift* of flowers on Lincoln's birthday.
You and Grindl should *interface* on that over lunch.	You and Grindl should *discuss* that over lunch.

ADDING SYLLABLES *Computate* is an invented word.

Incorrect	Correct
We'll use the calculator to computate the answer.	We'll use the calculator to compute the answer.
oscillitate	oscillate
nuculear	nuclear

COINING WORDS Leave it to the word mint, located in Denver.

Incorrect	Correct
I like to see a *proactive* attitude.	I like to see an *aggressive* attitude.
	I like to see a *forward-looking* attitude.
	I like people to *anticipate* needs.
	I like people to *act before being asked*.

There are so many correct choices because *proactive* is not a word. Therefore, the reader can choose whatever meaning makes the most sense at the time. Unfortunately, it may not be the meaning the writer intended. However, English evolves; no master sits on a mountain in Denver issuing new words and uses. *Proactive* seems destined to work its way into accepted use - just be careful that it says to your audience what you intend.

USING NUMBERS

Because most of us are in technical fields, we encounter numbers frequently. The rules for expressing numbers in written communication are: *be clear and be consistent.*

Numbers are either cardinal or ordinal. Cardinal numbers express a specific quantity, such as *one* or *three hundred*. Ordinal numbers express an order, such as *first* or *thirtieth*.

CARDINAL

Cardinal numbers express a quantity.

Cardinal Numbers

Spell out cardinal numbers less than 10, except for measurements of time, money, or dimension, in which case always use digits.

Incorrect	Correct
The chemical digester malfunctioned 3 times this week.	The chemical digester malfunctioned three times this week.
BUT	The chemical digester malfunctioned 13 times this week.

If a sentence contains cardinal numbers greater and less than 10, use digits.

Incorrect	Correct
The chemical digester must remain stable for three runs out of every 30.	The chemical digester must remain stable for 3 runs out of every 30.

Exception: ALWAYS spell out a cardinal number that begins a sentence.

Incorrect	Correct
13 stages of chemical digestion are required.	Thirteen stages of chemical digestion are required.

ORDINAL

Ordinal numbers express an order.

Ordinal Numbers

Spell out ordinal numbers below 10th

Incorrect	Correct
Correct grammar may not be the most important thing, but it's way ahead of whatever's in *2nd* place.	Correct grammar may not be the most important thing, but it's way ahead of whatever's in *second* place.
Our SuperSpell product is in its *twenty-third* revision.	Our SuperSpell product is in its *23rd* revision.

Exception: spell out an ordinal number that is part of a proper name.

Incorrect	Correct
James belongs to the *3rd* Augment of the Shining Knights of the Sea.	James belongs to the *Third* Augment of the Shining Knights of the Sea.

Exception: follow local custom for street names.

New York has *5th* Avenue, but Pittsburgh has *Fifth* Avenue.

FRACTIONS

Fractions have rules, too, but check with your company style guide. Technically oriented organizations, especially, insist on using figures for *all* fractions.

Fractions	
Spell out fractions less than one.	
Incorrect	**Correct**
Ceramics engineers always see the glass as 2/3 full.	Ceramics engineers always see the glass as *two-thirds* full.
Electrical engineers always see the glass as *one-and-seven-eighths* full.	Electrical engineers always see the glass as *1-7/8* full.
Exception: If a sentence contains fractions greater and less than 1, use digits.	
Big Company stock varied from *1/2* to *2-3/4* points today.	

TEMPERATURE

Not only must you worry about Fahrenheit or Celsius, but also about ° and degrees.

Temperature	
Incorrect	**Correct**
Set the furnace temperature to *650 degrees F*.	Set the furnace temperature to *650°F*.
When expressing a difference in temperature, follow the F or C with a degree mark.	
Incorrect	**Correct**
Increase the temperature by *100F degrees*.	Increase the temperature by *100F°*.

PARALLELISM

Parallelism means saying things in the same way when writing about related items. It helps to reduce unnecessary repetition and provides a way to add emphasis to your words. Also, it is a way of indicating equal rank among items.

Parallelism	
Perform the following steps to calibrate the instrument:	
Incorrect	**Correct**
1. Turn the power on. 2. The red light should be lit. 3. If the meter moves past 7, throw the Calibrate Switch to ON. 4. Before turning the Calibrate Switch ON, connect the Transducer.	1. Turn the Power Switch ON. 2. Make sure the red indicator light is lit. 3. Connect the Transducer. 4. Observe the meter; when it reads 7, turn the Calibrate Switch ON.
Every step in the corrected version is in the imperative mood. Therefore, the reader knows to expect an instruction of the form "Do this."	
Incorrect	**Correct**
The tests indicated a leaking gasket, a defective valve, and the temperature was too high.	The test indicated a leaking gasket, a defective valve, and excessive temperature.

Chapter 9 — Be Clear — Page 9-5

PRECISION

Precision is the mark of a technical person. A few ohms off and the circuit won't work; a little to the left, and the Holland Tunnel misses New Jersey. The same holds true in writing, especially in technical and business writing. You are writing for a very specific purpose; you must use the exact word you need.

Precision	
Don't say ...	... when you really mean
about	exactly
near	within two inches
many	most, some, all
none	few
all	most
person	Edgar Lynchpin (or other specific person)
terminal	workstation
CRT	monitor (monitors can be LCD or plasma, for example)
screen	monitor (the screen is only part of the entire monitor)
CPU	computer (the CPU is only part of the entire computer)
stress	strain
paper	report, manual, memo ...
liquid	molten
alternate	alternative

SPECIFICITY

Unless you are talking about abstract concepts intentionally, use specific words.

Concreteness	
Don't say ...	... when you really mean
Our company promises the highest quality products.	Our 90-day unconditional guarantee is proof of the quality and reliability of our products.
We need to improve quality.	We need to reduce the number of rejected units by 15 percent.
Our company is a good place to work.	Our benefit package ranks among the top three in our industry.

REDUNDANCY

Redundancy may be good to build into systems to make them more reliable, but redundancy in a sentence only makes the sentence wordy.

Redundancy

Don't say ...	... when you really mean
George asked for an early advance of his pay.	George asked for an advance of his pay. George asked to be paid early.
For correctly identifying the MIT Alma Mater, Ethel was awarded a free gift. (When was the last time you paid for a gift?)	For correctly identifying the MIT Alma Mater, Ethel was given an award.
Advance the tape forward, please. (What, advance it backward?)	Advance the tape, please. Run the tape forward, please.
After three days, the testing settled into a regular routine.	After three days, the testing became routine.
All throughout the experiment, the recorder behaved poorly.	Throughout the experiment, the recorder behaved poorly.
The end result of the program was to validate our conclusions.	The result of the program was to validate our conclusions. The program validated our conclusions.

SEPARATION

Modifiers (adjectives, adverbs, and their related phrases) need to be as close as possible to the words they modify. Otherwise, they become linguistic orphans, unsure of their real parents and set adrift in a hostile world.

Separation

Poor	Better
The transmission line, overloaded by the sudden traffic, began to sag.	Overloaded by the sudden traffic, the transmission line began to sag. The transmission line began to sag, overloaded by the sudden traffic.

Poor	Better
The circuit can be adjusted further to remove instabilities.	The circuit can be further adjusted to remove instabilities. The circuit can be adjusted to remove further instabilities.

The placement of *further* determines the meaning of the sentence.

Poor	Better
The engineer reported yesterday that the test had failed.	Yesterday, the engineer reported that the test had failed. The engineer reported that the test had failed yesterday. *(either, depending on the correct meaning)*

The last example is of ambivalent modifiers; they can't seem to make up their minds what nearby word to modify. Reword the sentence to help them decide.

PART 3: HELPERS

CHAPTER 10 — PAY ATTENTION TO THE FINE POINTS

Now you have seen most of the basic techniques for producing a clear, concise, professionally written communication. If you follow the rules, you would expect your writing to do the job you intend. However, because you write *to* and *for* other people, writing that merely is *technically* correct may not be good enough. Your readers also need to be comfortable with you, to think of you as someone they can trust and rely on. Reassure your readers by being respectful, polite, and honest. The basic rules of etiquette given below will help.

Also, because you write for business, you should respect the rights of the owners and operators of the business. Ways of making sure you do this are given below under Proprietary Information.

These are the fine points of writing, without which all your labors to build a functional and beautiful bridge between you and your audience will be diminished. Your bridge may carry the traffic, but it may not be an enjoyable journey.

ETIQUETTE

Etiquette may seem like a strange topic for a book on written communication, but consider what you do when writing: you ask for someone's time, energy, and possibly money. The least you can do is be polite. Being polite in writing is a little different from being polite in conversation or personal relationships, but the basic principles are the same.

RESPECT OTHERS

This means your readers. It is the key to being polite: if you respect someone, it will show. And you must respect your reader. After all, the reader is the one for whom you are going to all the trouble of writing. Consider it a privilege to be allowed to write for each reader. It is an imposition, after all, for you to expect people to take time to read your writing; make it a pleasant experience for them. They will appreciate it. The results from your audience analysis will help you to avoid topics and approaches that might seem disrespectful to your audience. For example, if your audience analysis revealed a secondary audience of state highway maintenance personnel, you would want to avoid jokes about and references to leaning on shovel handles.

BE POSITIVE

The tone with which you write discloses much about your feelings at the time. If you have had a tough day, or an argument with the boss, or if your computer ate the last 2 hours of work on that big spreadsheet, don't take it out on your reader. It is amazing what negative feelings can do to writing. Suddenly, things begin to be expressed in a negative sense rather than a positive one. For example, if you write, "The test did not produce acceptable results until the third trial," instead of "The test produced acceptable results on the third trial," you will convince your reader that the process is undependable, rather than demonstrating that it can be successful. Wait a while ... a few hours or a few days, if that's what it will take to get back to your sweet old self.

Do Unto Others...

Be helpful. Write the way you would like to have things written for you: clearly, concisely, accurately. Organize your writing to make it easy for the reader to understand and locate information. Minimize footnotes, and don't make the reader refer to the appendix every few pages. Make sure the page layout promotes understanding; keep illustrations and tables near the text that references them, and make those tables and illustrations clear, legible, and meaningful. Supply a meaningful table of contents, and maybe an index. Use page numbers that are logical and put them where the reader can find them easily.

Don't Tell Fibs

This is one of the most important rules. A reader who suspects you are not totally truthful, whether through commission of falsehood, omission of facts, or unidentified assumptions, will discount your conclusions and requests by a degree proportional to the dishonesty sensed. It takes only a little dishonesty to produce a lot of suspicion and turn a reader from a willing, cooperative partner in the communication into a total dropout or even an adversary. If that happens, your report is useless, your proposal a loser, your marketing brochure a waste of money. Don't try to fool your reader. If you have a good story, tell it; if not, explain why, and explain why it shouldn't matter. Usually, the reader will judge you fairly. And if you can't explain, maybe you should rethink why you are writing in the first place.

Don't Call People Names

Or companies. Or products. Readers want to know what you have accomplished, or are offering, or can tell them. They don't need or want your opinion on competing products or companies or processes. Certainly, they don't need your opinion on their own judgment. Some of the most annoying mail starts with a sentence like "Congratulations on making the wise decision to buy our fine product." Thanks, but I already figure it was a wise decision, since I made it. Why not, "Thank you for buying our product. We hope you are satisfied with it, and if not, please call our toll-free number listed below. We want you to be happy with your decision." Isn't that better?

DON'T BE STUCK-UP

Being stuck-up not only means thinking you are just a little better than everyone else, but it means showing that attitude. Your writing can seem stuck-up if it is pretentious, uses overblown language, or makes the reader feel somehow inferior. This last is easy to do, particularly if you use a lot of these words: *Obviously* (Oh, yeah? To whom?); *Intuitively obvious* (Is my intuition gone, too?); *Naturally* (Maybe I'm abnormal?); *As one would expect* (Gee, I didn't expect it!); or *As Figure 13 clearly shows* (So why isn't it clear to me?)

DON'T BE SELFISH

Share credit where credit is due, not only with coworkers but with other organizations within and outside your own company. If another department or division allowed you to use its facilities or personnel, say so ... and not only in a footnote. If the contributions are from another company or a university, or were published in a journal, you must acknowledge them and the organizations from which they came. If you are quoting or have used a trademarked product, you must make appropriate note. You cannot take credit for another's efforts, even if those efforts were peripheral to the main thrust of the project.

PROPRIETARY INFORMATION

In today's technical society, information often provides a competitive advantage. Consequently, companies have a proprietary interest in keeping some information confidential. It is your responsibility to ensure that confidentiality is maintained. And don't make the mistake of thinking that only technical information can be proprietary. Business plans and strategies, negotiation stances, discussions with labor unions, even company telephone directories frequently are considered to be proprietary information. Remember, proprietary information probably is not protected by copyright or patent, as copyrighting requires public disclosure, the very thing the company wants to avoid.

Most companies have procedures to deal with handling proprietary information. These procedures usually require a paragraph at the beginning of each classified document, and notation on each page. There may be a review procedure or a set of guidelines to determine one of several classifications. If you deal with the Department of Defense or other government agencies, strict adherence to classification levels, procedures, and control processes is a legal matter.

Not only must you protect company and government information, but you must protect the proprietary information of other organizations. Suppose your company has a temporary alliance with another for bidding on a particular contract for which neither alone has adequate expertise. Any information which your partner designates proprietary must be treated with the same precautions as your own.

If you receive bids from vendors in response to your own requests for proposals, you have an obligation to maintain the confidential nature of any information so specified. This is particularly true of pricing information, the disclosure of which could result in legal action.

The underlying principle in the safeguarding of information is "Think first!" before you write. Even though you may not consider a certain bit of data particularly sensitive, go through the clearing procedures. Maybe someone else knows something you do not. Better safe than sorry.

COPYRIGHTS

A copyright is exactly what its name implies: the exclusive or limited right of copying the form of expression of information. This means that copyright does not protect the information itself, but only the words or other means of communication used to express the information, often called the *work*. For example, if a description of an invention were copyrighted, a competitor would not be allowed to copy that *description*. But, the competitor could go right ahead and make and sell the invention if it were not patented. Copyright is important in those cases where the *expression is the product*. Examples include published papers, articles, and books; video and audio recordings; and some software, particularly the listing of the code.

The author of a work has entitlement to it (the right of copying it - the copyright) as soon as it is created. Strictly speaking, registering a copyright is not necessary; practically speaking, it is a good idea, as the following paragraphs explain.

It is necessary to claim copyright to a work only if you intend to publish it. The definition of publish is *to distribute copies to the general public*, and that could mean as few as one or two copies. There is some latitude concerning review copies and the like, but unless you intend to keep copies out of the public totally, you should copyright the work. It is the act of *public disclosure* that makes copyrighting necessary. Copyrighting puts the public on notice that you claim the right of copying the work, and anyone who wishes to copy it must get permission from you or face the possibility of legal action.

A copyright is unnecessary to protect a work only if the work is *unpublished*, that is if it is *not available to the public*. This means, however, that a company must copyright proprietary information that has limited distribution (called limited publication), such as would be the case with a proposal sent in response to a request for bid. However, there should be a notice of ownership and proprietary interest to protect the work from being disclosed to others. Some companies have taken to including both a proprietary notice and a copyright notice on such documents, although it is unclear if this really provides any additional protection.

Once a work is published, the only way to retain rights in it is to put on it a formal notice of copyright. The form for such notice is very specific according to both United States law and international agreement. It *must* contain the **copyright symbol** © (valid internationally) or the word **Copyright** or **Copr.** (valid only in the United States), the **year of first publication**, and the **name of the owner of the copyright** (the owner's logo or symbol is permissible if it identifies the owner unambiguously). It is not a bad idea to include the phrase *all rights reserved*, as that helps to legitimize your claim in certain Latin American Countries.

Note that I have not mentioned *registration* of the copyright. That's because registration or lack of it does not affect the validity of the copyright; registration only affects your rights as a copyright holder. If that sounds confusing, that's why there are copyright lawyers. If no one ever infringes on your copyright (that is, actually copies your copyrighted work without your permission), there is no harm in not having registered your copyright. Everyone is playing by the rules. However, if someone does infringe on your copyright, your rights to sue for enforcement of the copyright may be compromised unless you have registered the copyright. Got it?

Copyright registration is a simple procedure made complex by the variety of copyrightable material and by the fact that the Government runs the show. There are three basic steps to registering a copyright:

1. Fill out a form.

   That sounds simple, except that there are many different forms for different types of work. Get the wrong one and the registration is invalid, which you might not find out until you get to court with your infringer. Also, the law provides for stiff fines for falsely registering. Be a little careless, and it could cost you several hundred dollars. Get an attorney who specializes in copyright law.

2. Pay a $10.00 registration fee. Yes, only $10.00, believe it or not.

3. Send one or two copies of the work along with the first two items to the United States Copyright Office. Send one work if you are registering but have not and do not intend to publish the work. Send two copies if you have or will publish the work; one goes to the Library of Congress so the public really does have access to it.

In return, you will get a nice certificate to prove that you registered the copyright in case the official record is destroyed.

TRADEMARKS AND SERVICEMARKS

A *trademark* represents the owner's product; a *servicemark* represents the owner's service. It may be a symbol or a word or phrase, particularly if the word or phrase has been given a distinctive typographical treatment. Trademarks must be properly registered with the United States Patent and Trademark Office. Once a trademark is registered, others may not use it or even one similar enough to confuse consumers. However, trademarking does not prevent others from offering the exact goods or service under their own trademarks or servicemarks. It is *not* the goods or service that is protected: it is the *identification*. Be careful when writing about trademarked products or services. If you mention the trademark or servicemark of another, be sure to indicate that status. If the trademark or servicemark belongs to your company, be absolutely sure that the status is clearly indicated to protect yourself. Remember that cellophane and zipper were once trademarks that passed into the public domain because they were not adequately protected.

PATENTS

Patents exist to *encourage invention* by granting to the inventor the right to exclude others from making, using, or selling the invention for (under United States law) 17 years. As a condition for obtaining that right, the inventor must disclose the details of how to make and use the invention. The idea is that disclosure of this information in the published patent may give others inspiration to improve on the invention or to invent something different. The knowledge disclosed may also enable other inventors to avoid efforts that could duplicate the invention or that have been proven to be unsuccessful. Patents therefore serve to promote technological progress in two ways: by *rewarding inventors* with special advantages and by *making information available* to encourage other inventors.

For writers, the important thing to remember is that until a patent is *granted*, the invention is *not protected*; therefore it is vital that information relating to an invention for which a patent may be sought not be revealed to the public before a patent is obtained. To prove a patent claim, the inventor must keep working journals or technical notebooks documenting the progress in making the invention. Such records may be of fundamental importance should a court battle develop over patent rights. A court verdict may well depend on how clearly those records express the ideas the inventor followed, if they establish certain dates, and on whether or not the record was witnessed.

TRADE SECRETS

Trade secrets are neither patented nor copyrighted, because they are so sensitive that disclosure of them in a public forum could place the company at a competitive disadvantage. Therefore, they must be protected with great care. Perhaps the competition doesn't even know a secret exists! A trade secret may be a process, a formula, or a technique used to make a more desirable product. However, just because it is not copyrighted or patented does not mean there is no protection. For example, an employee who disclosed trade secrets to a competitor might be prosecuted under criminal or civil law, particularly if the employee had been in a position of trust. If a competitor discovers a trade secret through its own research or even through reverse-engineering, the secret is, by definition, no longer secret, and no protection exists.

CHAPTER 11

GET THE COMPUTER TO HELP

This is the age of the computer, no doubt about it. And for the writers of business and technical communication, that is good news, because computers can help in many ways. Computers can help you with spelling, style, grammar, organizing, rewriting ... even with thinking! And as a writer, you should not hesitate to use these computerized assistants whenever you can, as long as you realize their limitations - which are sometimes considerable.

WRITING ASSISTANTS

Writing is hard work. It is also work that requires more than a little ego to do well. After all, you are assuming that what you have to say is interesting enough for other people to take the time to read. Consequently, you owe it to both your readers and your ego to make sure that it is. You could just sit down with a pad of paper and a pencil (and many successful writers do exactly that), but computers have made the writing process easier and more efficient.

The most common manifestation of this assistance is the word processor. Most technical professionals have access to a word processor of some sort, and the general feeling about writing on a computer is that it is a liberating experience. The ability to easily change, delete, and add words and paragraphs encourages revisions that might otherwise be neglected. Consequently, writing tends to be better and tighter. Unfortunately, there is a also a tendency toward the opposite: it is so easy to write in an almost stream-of-consciousness style that writing may tend to ramble, to be wordy and redundant. A good writer will take the time to review, revise, and rewrite until the document is concise, accurate, and complete.

The computer also provides tools to make other aspects of writing easier. Outliners and thought processors help with brainstorming and organizing; electronic thesauruses provide variety; CD-ROM players and online data bases provide nearly unlimited research facilities; and hypertext systems allow writing in a nontraditional, nonlinear manner.

Each of these tools has a place, and each can be of considerable help. But none can substitute for proper analysis of requirements and audience, or for diligent and conscientious application of the principles of good writing.

OUTLINER

An outliner allows the writer to write without much thought to all the rules of grammar and punctuation. An outliner can let you brainstorm at the keyboard as in this example.

Suppose you are writing about a new idea for a product that will revolutionize life for cat owners. You might have the following ideas which would look like this if you used an outliner:

a) less expensive

b) cats like it

c) less dirt

d) no smell

e) keeps away dogs

f) your family will like it

g) it would be a great gift

h) better than catnip

i) buy anywhere

j) recommended by veterinarians

An outliner lets you re-order these thoughts with just a few keystrokes or movements of the mouse.

a) cats like it

b) better than catnip

c) keeps away dogs

d) your family will like it

e) less dirt

f) no smell

g) it would be a great gift

h) less expensive

i) buy anywhere

j) recommended by veterinarians

Notice that the outline has been reordered automatically.

Then, you can make some subjects subordinate to others and add text as required. Most outliners allow you to hide text and subordinate levels to reduce confusion. This keeps the logical structure of your developing document clearly in focus.

a) cats like it

    1) better than catnip

    2) keeps away dogs

d) your family will like it
   1) less dirt
   2) no smell
g) it would be a great gift
   1) less expensive
   2) buy anywhere
j) recommended by veterinarians
   1) endorsed by the League for Animal Happiness.
   2) supplies needed vitamins

Finally, most outliners can export the outline to a word processing format for final polishing and printing.

THESAURUS

An electronic thesaurus is much like the book you had during freshman English (or should have had). By selecting a word and hitting a hot key, you call forth synonyms, sometimes with their meanings, so that you can select the word with the exact shade of meaning you intend. Usually, each of these synonyms can be subjected to the same look-up procedure, resulting in yet another set, and so on until the list begins to repeat. It is not uncommon for a good thesaurus to list dozens of synonyms for the original word. If word meanings are given, it makes your job even easier. Using an electronic thesaurus can give precision and variety to your writing, but overuse can lead to stilted language. Remember to use familiar words. Don't plug in exotic synonyms just because they look interesting. Stick to the rules: each word has a purpose, so choose the word that best fulfills that purpose.

THOUGHT PROCESSOR

Sounds a little like a mental Cuisinart (TM)! Thought processors (or idea generators, or problem analyzers) are intended to liberate your thinking, to help you remove barriers of conventional thinking that may suppress your best ideas. Thought processors are like outliners in that most allow you to capture random thoughts. But thought processors go another step. You can assign one or more categories (or other identifier) to these random thoughts, so you can use the manipulative abilities of the program to discover new associations between them. In this way, you are forced to look at the relationships among your thoughts. This in turn often results in new thoughts, new categories, new insights, and the process repeats until a clear and original approach emerges. Your writing benefits from the freshness of thought; you are forced to explain new and unfamiliar concepts; you are stretched, and so improve.

Some thought processors go even further, engaging the user in a dialogue to stimulate creative thought. For example, the program may ask: Is this a problem concerned with process, procedures, or communication? What conditions or actions will cause this process to fail? What are the consequences of failure?

The program can't give any answers, but it can make sure you have covered most of the areas needed for effective problem solving ... and for most of us, that's a big plus.

HYPERTEXT AND HYPERMEDIA

Hypertext makes possible a document that may be read in nearly unlimited ways. This means that the reader can locate information quickly from nearly any point in the document, and follow trains of thought that are not what the writer had in mind. In other words, instead of the linear organization of the standard document, imposed by its outline, the hypertext document has many different potential organizations, which are a function of the number of permutations and combinations of the links between information *chunks*. Further, the reader may return easily to the starting point, regardless of the number of intermediate levels or cross-references through which the train of thought has passed.

The advantage for the writer is that the document can be written in modules, with mini-documents describing discrete chunks of information and with nearly unlimited links connecting these chunks to each other with nearly unlimited variety.

Hypertext requires the writer to decide exactly what chunks the information should be broken into. This imposes a very disciplined approach, which can only improve the final result. The disadvantage is that the links, too, are defined by the writer; the reader does not have the *ideal* unlimited variety of access, but is restricted to those pathways the writer decided were worthwhile. Although this limitation in some ways hinders the creative aspect of the hypertext concept, as it may suppress some mental links the reader may have made but have been unable to follow, hypertext still can surprise when a reader manages to pursue some imaginative combination of links to a new conlusion or revelation.

The practical implementation of hypertext involves the writer identifying words, phrases, identifiers, graphics, or other elements within the hypertext document as buttons or references. When the reader activates one of these buttons, the link calls to the screen the linked text or other object. In this way, an instruction manual might have buttons on component names which link to part numbers, and then to a drawing of the component, and from part of the drawing to instructions for replacing that particular subcomponent.

The newest variation on hypertext is hypermedia. This is the logical extension of the static text and graphic hypertext system which enables the author to create links to such interesting items as video displays, audio recordings, automatic logons to remote data, or information retrieval systems.

CD-ROM REFERENCE

An accelerating trend is toward data storage media of greater and greater capacity. The current leader in this field is the CD-ROM, or Compact Disc-Read Only Memory. This is almost the same device as the audio CD player that is displacing audio cassettes for music lovers. In fact, if you aren't using your CD-ROM to help you write, you probably can use it to listen to your favorite music.

The main advantage to CD-ROMs is their enormous capacity: they start at around 600MB, as compared to 360KB to 1.44MB for most floppy diskettes and 40MB to 200MB for most rigid magnetic disk drives.

So what do you need all this room for? How about the entire text of an encyclopedia? Illustrated. In color. Indexed and cross referenced. Able to retrieve any information using logical search keys as quickly as within one-third of a second.

Or an almanac. Or the ZIP code directory. Or any of a dozen specialized dictionaries (foreign language, technical, medical, and more).

The possibilities are endless, and more and more offerings are coming out daily. Some periodicals are becoming available on CD-ROM disks. So what's the drawback? Cost, for one thing. A CD-ROM drive costs around $500 at the time this is being written, and CD-ROM disks go for between $100 and $300 each. That's quite an investment to replace your old dog-eared Britannica. But have heart: at the rate technology is proceeding, these devices probably will be standard equipment for the well-equipped writer within a few years.

EDITING ASSISTANTS

By definition, editing happens after writing. Editing ensures that writing meets standards of grammar, spelling, punctuation, and readability. Editing ensures that language and style are appropriate for the audience and subject and that the document is well organized and complete. Although no computer program (yet) does all that, there are several tools you can use to make the editing task easier.

Spelling Checker

The news about spelling checkers is both good and bad. The good news is that more and more word processing and desktop publishing programs include spell checking as an integral part of the program. More good news is that these spell checkers get better and faster with each new release. The bad news is that most of them only check spelling ... which means that making sure you have the right word is still up to you. A spell checker can tell if *persuade* is right or wrong, but cannot tell if *toe* should be *two*, or *too*, or *to*. That is still up to the writer. Also, consider this: your trusty spell checker detects a misspelled word and offers three alternatives; which one is correct? It may not be immediately obvious, depending on the word and the meaning you intend.

The following paragraph was passed without error by three spell checkers:

Deer Sur,

Eye halve wonted two right too ewe. In reverence too yore Spell check soft wear. Eye red threw most awl their Liszt end its reel awed butt Eye kneaded two beyond aye hire plain two under stand.

Grammar Checker

Also called style checkers, these programs are intended to alert the writer to incorrect grammar and inappropriate style. After having gone through the earlier chapters of this book on grammar and style, you probably heartily welcome such programs! However, as with spelling checkers, there is both good and bad news. The good news is that these programs exist and are getting better all the time. The bad news is that their reach exceeds their grasp.

While most of these programs do check grammar at some level, the complexities of the English language are their undoing. Grammar checkers, like spell checkers, are fine as far as they go, but they are no substitute for an informed writer. The writer must decide if the suggestions the program makes are applicable, or (unfortunately) in many cases, whether the suggestion is simply right or wrong.

Let's look at what grammar checkers can do.

Most offer suggestions to check on some particular point of grammar, or question if the language is too complex or vague;

others look out for unnecessary or incorrect punctuation, or inappropriate words.

Almost all calculate readability, usually expressed by the grade level needed to comprehend the writing. Most report a measure of strength (simple writing rates as strong), and most supply warnings of excessive use of jargon.

Some grammar checkers allow users to specify the type of writing: technical, fiction, advertising, and so on. Then, the program applies different rules according to the type.

Many programs cannot tell whether *their* or *they're* is correct in context, and so alert the user to every instance of either word, while others are intelligent enough to recommend the appropriate correction in most cases. And that's the problem: in most cases.

Do yourself a favor. Use a grammar checker on your writing as you would use a ripsaw on fine furniture stock: for the rough cut. Depend on your own craftsmanship to supply the fine work.

REDLINER

Redliners are to word processors what "what if" routines are to spreadsheets: they let you explore alternatives without committing to changes. Named for the red pens frequently used by reviewers to mark comments, redliners allow others to suggest changes and insert comments directly into your text file. Then, you can review the suggestions and comments and choose which to accept or reject.

On networked systems, this redlining process can involve several reviewers and several revision cycles quite productively. In simpler situations, redlining can be an effective tool even for editing your own writing, giving you the chance to try different options without actually changing the file (and then having to change it back later).

If your word processor does not support redlining, you can improvise by enclosing suggested deletions in brackets and typing suggested additions and changes in all caps, for example.

DOCUMENT COMPARER

These handy programs compare the text of two similar word processing files and report any differences. This is useful in comparing revisions to the same document, or comparing members of a series of tailored documents all based on the same original file. Comparing documents to a common base document quickly points out, for example, unusual items in a contract, due dates, costs, quantities, or addresses that differ from the base. Comparing documents is helpful if several independent reviewers alter copies of the original file, or if you have forgotten which version of a document you really want.

MANAGING THE WRITING PROCESS

No matter how many pages your document includes, or how simple or complex it is, or if the audience is the vice president or your office mate, you can use all the help you can get to keep the details straight and make the finished product look good. The following techniques and tools can help.

FILING SYSTEM

It is always a good idea to be organized, especially when writing a technical or business document. Organization helps you to keep on schedule, to make sure you have all the needed information, and to know the status of each part of the project. Checklists and filing systems can be made easily and quickly using word processors, data base managers, and spreadsheets.

Many word processors can generate an index and table of contents. If you update these as you proceed through the various drafts, a quick look at them will tell you what parts are missing. An increasing number of word processors allow the user to append notes anywhere in the text, which can be hidden or disclosed at the user's option. You can use these notes to remind yourself of missing information, to collect pertinent data, to document a source, or to note the due and completion dates for chapters or sections.

A simple data base will let you collect and organize research notes and bibliographical information. If you organize the database well, you will be able to use it to sort your data in any way you wish. Further, most data base programs can create custom reports of the data, and export that data into a file readable by most word processing programs. This allows you to easily transfer data from your research notes into the finished document.

A spreadsheet may seem like an unusual tool to help organize writing: most people think of spreadsheets as tools to calculate numbers. But if you think of a spreadsheet as a large variable table, you can find many ways to make it useful. For instance, list each section in the first column, the estimated number of pages in the second, the author (if you are collaborating) in the third, the due date for the first draft in the fourth, and so on. Then, you can sort the spreadsheet on any due date and see exactly where you are in the process, and which sections are behind, on time, or ahead of schedule. This will allow you to decide where your time would be spent best in order to meet the final publication date.

These are only a few of the ways in which ingenuity and some common software can help you get control of the main steps involved in the writing and publication processes. The more you experiment, the more ways you will find that work for you. Be creative!

STYLE SHEETS

Style sheets used to be the sole property of the typesetter and editor. Together, they used them to decide how to rearrange your writing so that it conformed to the typographical standards for the document you were writing. Now, much of that work can be performed by word processing programs or page-makeup programs (also known as desktop publishing systems).

An electronic style sheet allows the document designer to specify the typographical elements (the individual styles) of standard parts of the document. The style sheet is the collection of these individual styles and it applies to a particular type of document. For example, a manual might have a style defined for numbered step procedures, one for unnumbered step procedures, one for cautions, and another for warnings, as well as one for the general body text. Then, either while writing or later, a simple assignment of a style to a section of the document will produce the desired appearance.

Electronic style sheets save time and result in a consistent document that helps the reader. All the typographical features of this book resulted from the application of styles.

CHAPTER 12   WRITE!

Now it is time to take the big step ... to actually *write* something. As you have seen, writing is more than putting words down on paper or onto a magnetic disk; writing comprises all the tasks covered in the previous chapters of this book, from figuring out just what it is you need to write to picking the computer tools to help produce your document. Those steps are repeated here, all in one place, so you can get a grasp of the whole of which so far you have seen only parts.

STEP 1: DEFINE THE JOB

Accurately defining what the job really is will save you considerable effort. Drastic rewrites and edits can be avoided by doing the right thing. This means making sure you know the *audience* and the *purpose*.

DEFINE YOUR AUDIENCE

Complete the Audience Analysis Checklist. Make sure you have determined:

Audience type
- Who is the primary audience?
- Who is the secondary audience?

Audience complexity

Audience viewpoints

Audience familiarity

Audience needs

DEFINE YOUR PURPOSE

Complete the Writer Analysis Checklist. Make sure you have evaluated why you are writing by defining the following:

Category
- Primary category (inform, request, persuade)
- Secondary categories

Reasons (within each category)
- Primary reason
- Secondary reasons

Goals
- Primary goal
- Secondary goals
- Hidden agenda items

STEP 2: GET READY

Writing comes much easier if you are ready when you begin. Make sure you know the basic structure of your document, have all the information you need, know when things need to be done, and know where to go for the help and other resources you might need.

GET ORGANIZED

Consider all the techniques listed in Chapter 5 to help you get a clear picture of how you need to organize your document.

ANALYZE AND COMPARTMENT-ALIZE

» What information must you include?
» Must you follow a set outline?
» Is there a page number limit?

OUTLINE, WRITE INSIDE-OUT

» Is the order logical?
» Are the right topics subordinated correctly?
» Does the structure accomplish your goals?
» Is the structure easy to follow?
» Does the structure conform to any imposed requirements?
» Have you *Told 'em* ... ?

USE VISUAL CLUES

» What can you do to help the reader grasp your ideas?
» What will your document look like?
» Can you choose the format, or has it been established?
» What messages do you want the format to send?
» Do you have enough time, money, and expertise?

INCLUDE SUPPORTING DATA

» Do the data support the conclusions?
» Are there enough data to justify the analysis?
» Is the presentation of the data clear?
» Have you put in adequate references?

SCHEDULE THE WORK

As mundane as this seems, it is important to do or to ensure that someone else does. After all, the best document in the world is of no use if it arrives after the deadline.

» When do you need to be done?
» What parts will take the most time?
» Where will the sticking points be?
» How much time is needed for review?
» How much time is needed for writing?
» How much time is needed for editing?
» How much time is needed for producing graphics?
» How much time is needed for page layout?
» How much time is needed for printing and binding?
» How much time is needed for shipping?
» Prepare a detailed schedule and stick to it.

TRACK DOWN RESOURCES	A lot of details need attention before you can translate your ideas into printed pages. If you look into them now, you may avoid grief later on when, for example, you find that the word processing program you used cannot be translated into the company's publishing system. More important are the people on whom you must rely. If you can't get people to cooperate, or if you don't know whom to contact for that critical piece of information, you could be in a difficult position. And speaking of data, you should make sure that the data you are planning to incorporate are available, and that you will be permitted to use them. You wouldn't want to have to spend a lot of time and money on generating new data if another approach to the document could avoid it. Below are some things to think about when considering equipment, people, and data:
EQUIPMENT	» What word processing equipment is available? » Where is it? » Do you have access to it? » How much keyboarding help can you get? » What graphics equipment or programs are available? » What reproduction (printing) facilities are available?
PEOPLE	» Who has the information you need? » When can you see them? » Will they cooperate? » Do you need to give them anything in return? » What stake do they have in the document? » Do they have time? » Are they allowed to help?
DATA	» Where is the information you need? » What form is it in? » How accurate is it? » How complete is it? » How long will it take you to use it? » Can you get it? » What authorization will you need to get it? » What analysis will need to be done? » Can you do it, or must you get someone to help? » What information is missing and how can you get it?
STEP 3: WRITE	This should be the easy part, now that everything is in place. Just make sure that you follow the organization you have prepared; obey the standards of grammar, punctuation, and spelling; value clarity and completeness above all; and take care not

to break any company, legal, or personal rules. Remember that there are computerized tools to help you in all of this, but don't rely on them blindly.

USE THE RIGHT STYLE
» Are the nouns and verbs in agreement?
» Are there double words?
» Have you checked for danglers of every sort?
» Are the sentences *really* sentences?
» Have you used the appropriate vocabulary?
» Is the writing bias-free?

BE CLEAR
» Have you used numbers correctly and consistently?
» Have you used only *real* words, and avoided verbizing?
» Are the verbs out in the open and not hidden?
» Have you used mainly active voice?
» Do you state any assumptions clearly?
» Are your arguments and lists parallel?
» Have you eliminated redundancy?
» Is your language precise and unambiguous?
» Have you kept modifiers and the modified together?

PAY ATTENTION TO THE FINE POINTS
» Have you followed the rules of etiquette?
» Have you given credit where due?
» Is any of the information proprietary?
» Have you authorization to use company information?
» Have you permission to use copyrights, trademarks, and service marks?

GET THE COMPUTER TO HELP
» Have you used a spelling checker, and double-checked it?
» Have you used a grammar checker, and triple-checked it?
» If you used a redliner, have all comments and redlines been addressed and removed?

STEP 4: REWRITE
This may be the most important part of writing. Many writers believe this is where the real work is. It is also where your ego is on the line, because you must take a critical, detached look at what you have written.

REVIEWING
If you can, put your document aside for a few days after it is completed. Then, take a hard look at it from the reader's viewpoint. Is it clear? Is it easy to follow? Is it literate? What about typos, inconsistencies in punctuation, acronyms, and nomenclature? Next, look at it from your own and your company's viewpoint. Does it do what you want it to do? Does it meet all of the stated and hidden agendas? Does it make you

and your company look good? Does it say **competence and quality**? Unless you are a very unusual and talented writer, the answer to some of these questions will be *No*, no matter how well you prepared. Writing is just too tough and complex to get it *exactly* right the first time through. It might not be a bad idea to ask a colleague to review the document, too.

REVISING, REWRITING, AND EDITING

Now that you have reviewed your document and know where all the rough spots are, take the time to *revise it carefully*. Make sure that the revisions not only satisfy the concerns expressed during review, but that they meet the standards you applied to the original work. Then, after all the changes are made, review it *again*. And *again*. And maybe *again*. Remember, this is how poor to mediocre writing becomes really good: in the rewriting. Besides, if you did a really good job with Getting Ready to Write and Writing, the rewriting should not be excruciatingly painful. Further, you will be surprised to notice how much better your original writing becomes after you have rewritten major sections.

If you are part of an organization that has a staff of professional editors, and if your budget allows, take advantage of that capability. The value added by a person whose business it is to ensure that documents work cannot be overestimated. And don't fear that your golden prose will be destroyed or the meaning of your writing will be changed: these people are professionals, and won't let that happen. They simply will help you make your document the best it possibly can be.

STEP 5: PUBLISH

A dusty, muddy Mercedes-Benz is still a Mercedes-Benz, but it just can't compare with one that has been washed and polished, and sits gleaming in the sun. The same is true of a document. A document that is poorly reproduced and bound, or arrives damaged because of improper shipping, loses something. The reader wonders how much you really care, and that can change the reader' perception of the entire document. You've spent a lot of time writing this thing; don't try to cut corners now.

» If you are part of a large organization that has a separate group for reproduction, binding, and distributing, let them do those things. They know the best ways of getting these tasks done, and will do so in a way that presents your document in the best possible light. Besides, you have other things to do ... like getting on to your next writing task!

» If you must do these things yourself, ask the secretarial staff if there are standard procedures for copying (one side or two), for binding (loose leaf, plastic comb, staple ...), and for distributing (hand carry, internal or external mail, express ...).

STEP 6: RELAX

You've done all you can to ensure that your messages get through. Now it's up to the reader to decide if it was enough. Fortunately, if you have applied the techniques presented in this book, you will have improved your chances of success considerably. Unfortunately, life is uncertain and there are no guarantees, so you might have missed the mark. If so, don't feel bad; even professional writers have a bad day. What is important is that you *keep practicing* the principles you have learned. They work, and you will be a better writer and a more successful professional for the effort. Further, you will have become a better communicator in general, and that's a valuable trait in this increasingly complex world. Ideas are the stuff dreams are made of, but dreams won't be realized unless the ideas are communicated.